T0276025

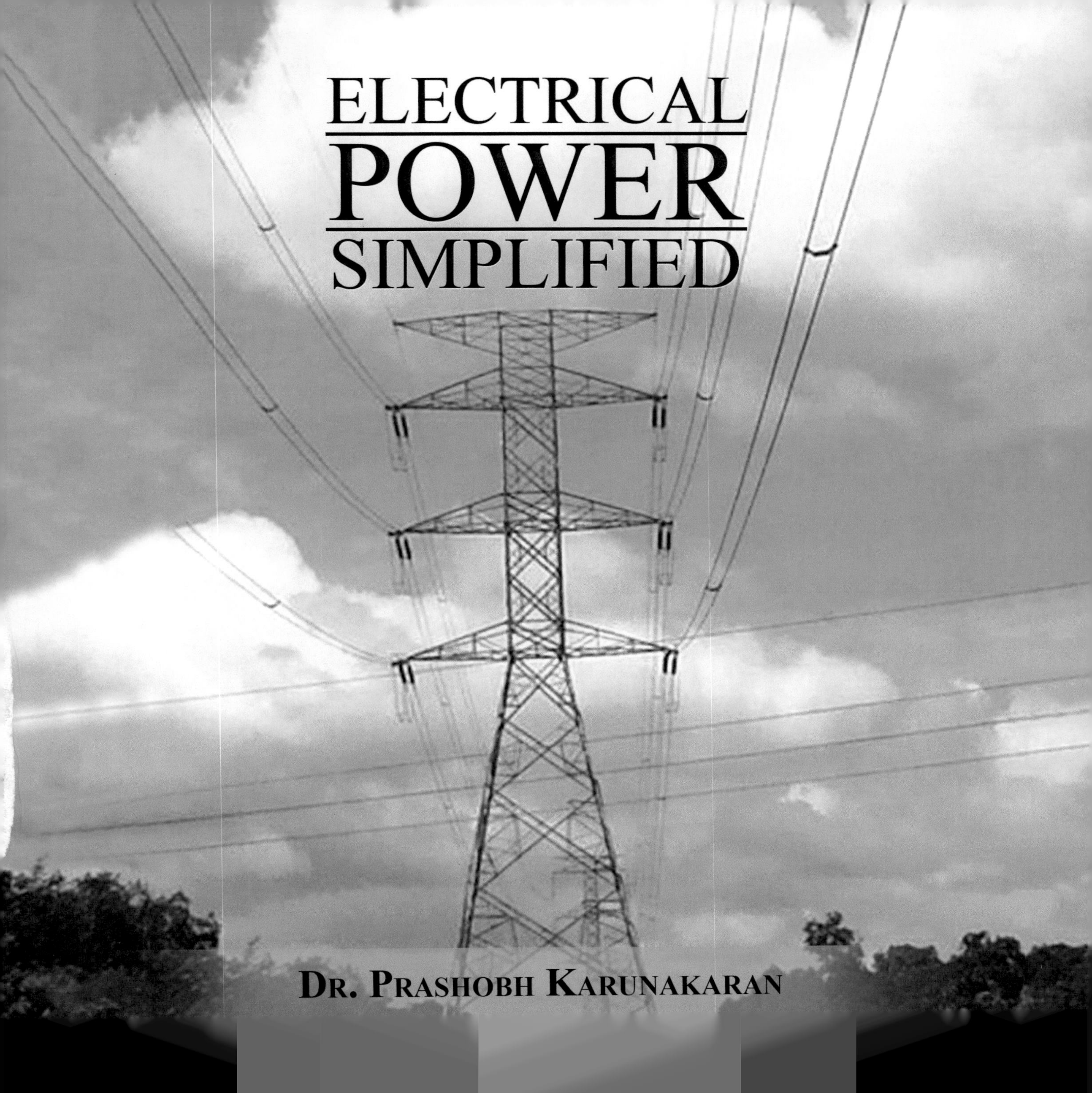
ELECTRICAL
POWER
SIMPLIFIED
DR. PRASHOBH KARUNAKARAN

*AuthorHouse™*
*1663 Liberty Drive*
*Bloomington, IN 47403*
*www.authorhouse.com*
*Phone: 1 (800) 839-8640*



*Published by AuthorHouse 10/10/2018*

*ISBN: 978-1-5462-6246-6 (sc)*
*ISBN: 978-1-5462-6247-3 (e)*

*Library of Congress Control Number: 2018911740*

*Print information available on the last page.*



authorHOUSE®

# ACKNOWLEDGEMENT

Special thanks must be given to Edi Jungan, Harry Ante, Mid Dewang, Phang Su Ling and Ganesa Ramamoorthy for their help in getting some of the materials for this book. Great thanks must also go to my wife, Sreeja and children, Prashanth, Shanthi and Arjun for supporting in the writing of this book.

# PREFACE

This book simplifies electrical power engineering and provides a working knowledge of the field. Equations are avoided as far as possible. I did a Bachelors in Electrical Engineering at South Dakota State University, SD USA and then switched to a Masters in Economics; mainly because the Engineering Department required me to pay fees. The Economics Department offered me full scholarship since I became the top out of 80 students in two economic courses I took in the summer break after securing my engineering degree. I continued acing the Masters subjects and did better than Rich who shared an office room with me at the University. But I always knew Rich was much better than me at Economics. I aced economics because of my engineering knowledge of mathematics and graphs. I needed the tools of equations and graphs to explain things while he could just talk about economic principles fluidly. Of course, I have since moved back to engineering and spend the last 26 years striving to reach the fluid knowledge of electrical engineering that Rich had for economics; including doing a PhD in engineering. I can now talk about electrical engineering as Rich talked about economics, without referring to the tools. This book is what bloomed out of that knowledge.

# Contents

# Chapter 1

# Introduction

Electricity has provided huge benefits for mankind. The small village homes depend on it for lighting and irrigation of crop lands. At the other end of the spectrum, the prime mover of the largest machines in the world have over the past few decades moved away from combustion engines and hydraulics to electric induction motors. Extremely hot areas of the earth like the Arabian Peninsula have been made habitable with electricity powered air conditioning and extremely cold places like Alaska can attract human populations with electricity powered heating. Thus, electricity has become a fine compatriot of humanity. But, if we touch an electricity carrying wire, we will be burnt. It is therefore imperative that as many people as possible have proper knowledge of the limits, dangers of electricity and respect it. This book hopes to disseminate knowledge of the electric power system to as many people as possible. This is especially so because it is beginning to power an ever-increasing number of human activities which were formerly powered by combustion and hydraulic systems. The largest cranes, largest ships and largest trucks in mines are already electric powered. The predictions are that electric cars will replace combustion engines ones seems easy to understand, because non-engineering people in politics (the vast majority) in countries like China, India, Britain and France and Indonesia have already set target dates for all cars to be electric motor driven. Calculations are avoided as far as possible in this book. The most useful formulae in electrical power are Ohm's law (1) and its derivation, the power law (2)

$$V=IR \quad (1)$$
$$P=VI \quad (2)$$

Where V = Voltage or EMF (Electromotive Force) in volt units, I= current in ampere units and R = resistance in ohm units and P = power in watt units. Other than the Ohm's law another fundamental thing to remember in electric circuits is that current is the same over all components connected in series and voltage is the same over all components connected in parallel. Simply put **current is the same in series** and **voltage is the same in parallel**.

The simple principle of electricity is that generation of power should always equal to the customer demand for it. If this is not balanced, there will be effects on voltage and frequency. The control of this balance is the biggest complication in an AC (alternating current) electrical power system. Nowadays intermittent renewal energy is slowly replacing conventional hydrocarbon energy which has suddenly made the system even more complicated.

Propagation speed of electricity is affected by insulation. In an unshielded copper conductor, it is about 96% of the speed of light, while in a typical insulated coaxial cable it is about 66% of the speed of light ($3 \times 10^8$ m/s). But actual speed of electrons is near 0 m/s in AC Alternating Current) and comparable to putty flow down a wall in DC (Direct Current). An analogy to explain this is, if a pipe with ends named A and B are filled with table tennis balls. And if the first ball at point A is given a little push, immediately the ball at end B will move. In a similar way, a push of the first slow moving electron at point A of an electrical wire will cause the electron at point B of the wire to move immediately; this immediate action is termed current (I). This action happens at almost the speed of light in a bare electrical wire (conductor). If the conductor is covered with an insulation, current flow can drop up to 66% the speed of light.

For current carrying wire, the analogy of the finger which pushed the first ball in the pipe is replaced by a term called voltage or EMF (Electromotive Force). Thereby voltage is the force which pushed the first electron giving a force to each and every ball all along the pipe. If the finger pushed the first ball with a great force at end A, the tennis ball at point B will jump out and fly quite a distance. The combined effects of the finger force plus how many balls are passing one particular point of the wire at one span of time, is termed the power (so P=VI, whose units is watts). If the pipe is rough (high R), even a forceful push (high V) of the ball at point A will not result in the ball at point B moving much. But if the pipe is made as smooth as humanly possible (low as possible R) even a slight finger push (small V) will cause the ball at point B to move with close to exactly the finger pushing force exerted at point A.

Another analogy is a powerful BMW next to a tiny Daihatsu on a straight road. The BMW has high power (high EMF or V) and the tiny Daihatsu has low power (low EMF or V). The speed and size (quantity of electrons) of the car is the current (I); sometimes the tiny Daihatsu can actually beat the powerful BMW if the driver is good enough, that is, the current of the Daihatsu can be greater than the BMW even though it has a low V. When the two cars meet a traffic jam, they are experiencing R (resistance).

The total resistance to electron flow is actually termed impedance or Z where $Z^2=R^2+(X_L-X_C)^2$ because other than the traffic jam mentioned above, there are two other ways electron can be slowed:

1) **Resistance:** the 'roughness in the pipe' is termed resistance (R). A rough pipe will restrict the string of table tennis balls from moving or flowing which is equivalent to resistance. Because the pipe is rough, the balls have to vibrate a lot more and the higher the vibration lateral to the pipe of course the slower the linear speed through the pipe. Factors that influence the vibration (or resistance) are classified as $C_a$, $C_g$, $C_i$, $V_d$ and cable table. $C_{\mathbf{a}}$ represents the **a**mbient temperature. $C_{\mathbf{g}}$ represents the **g**rouping of the number of cables running in a particular pipe. $C_{\mathbf{i}}$ represents the roughness of the pipe (type of **i**nsulator used) $V_{\mathbf{d}}$ or voltage **d**rop, represents the length of the wire and the cable table represents the carrying capacity according to the cross-sectional area of the conductor.
2) **Inductive reactance:** electrons can be equated to small magnets, which is why a compass is deflected when placed next to a DC current carrying wire as was discovered by Hans Christian Ørsted. Later André-Marie Ampère discovered that electrons moving in a circle parallel to each other (as in a solenoid) forms a magnet. His discovery of the solenoid was later turned into a practical instrument by Joseph Henry. Assuming the table tennis balls in a pipe are magnetic balls and the pipe which is bent till it becomes a coil. Because of the coil, the magnetic balls are concentrated in a small region which equates to a strong magnet. This concentration of magnetic field in a small region of the wire results in all the balls having to move slower within this stronger magnetic region (solenoid). Also, the balls must travel a longer distance as they move in coils just as cars need to move slower in a roundabout compared to a straight road. Therefore, the speed of the magnetic balls is further reduced. The slowing of electron flow (current flow) as it moves into a solenoid is called inductive reactance. This is the same as placing a permanent magnet over a current carrying wire. In a permanent magnet, free electrons from iron (or other ferromagnetic elements) spin parallel to each other (just as in a solenoid), within regions called domains making each domain a tiny magnet. If these tiny magnetic domains are all aligned, the iron becomes a magnet.
3) **Capacitive reactance:** assuming the negatively charged table-tennis balls have built up on a metal plate and another plate is nearby connected to the negative electrode. The second plate is neutral (having an equal number negatively charged tennis balls and stationary positive charged balls). If more and more of negatively charged (and movable balls) accumulate on the first plate, a critical charge will be reached whereby these balls can have enough energy to jump and escape to the next plate which is

positively charged. This accumulation of electrons on one plate before jumping onto the second plate is effectively a delay in electron flow called capacitive reactance.

All three; resistance, inductive reactance or capacitive reactance slows down electron flow in different ways summarily by 1) vibrating a lot, 2) being in a magnetic field and 3) by having to jump across a small gap.

# Chapter 2

# Electrical Safety

We cannot say current or voltage is more dangerous. The combination of both, the power whose equation is below is what causes electric shocks.

$$P=VI \cos \theta \quad (3)$$

Where θ or cos θ is a measure of the shifting of the current waveform with respect to the voltage waveform. This shifting in the travelling speed of the current waveform with respect to the voltage waveform is called phase shift. The phase shift only occurs in AC moving through a capacitor or inductor and doesn't happen if the AC is moving through a resistor. Phase shift also does not happen in DC.

It takes about 40 volts of force for the electrons from a current carrying conductor to jump into a human skin. These 40 volts, plus or minus a little will form a sigmoid curve if human resistivity data is collected. People with moist hands will conduct electricity slightly better than those with dry hands (note moist hands include salt, pure water does not conduct electricity). An equipment can have very high power but insufficient voltage for that current to jump into your skin. For example, a very big and high-powered speaker can have 12V leads but high current (amps) to deliver such huge power to the speakers seen in some dance places. A human can touch this 12V lead even when the speaker is live.

Definition of danger in electrical installation is, any source of voltage which is high enough to cause sufficient current flow (electron flow in reverse) to the muscles or hair. The real thing that is moving in electrical flow is electrons but due to the mistake of Benjamin Franklin we are all using the direction opposite of electron flow which is named current. If calculations are made with consistently taking opposite direction of the actual electron flow, the results will remain the same.

Electric currents are always finding a pathway to go to Ground (Earth). If there is an easier pathway for it to reach Ground (Earth) via a copper wire, it will avoid choosing human bodies to reach ground. The ground resistance (earth resistance) must be <100Ω for homes, offices and factories, it must be <10Ω for a steel electric poles and <1Ω for substations and power stations. To put into perspective why electricity loves to go to the ground; if two 10Ω resistors are connected in parallel, the resultant resistance is 5Ω. If four 10Ω resistors are connected in parallel the resistance drops to 2.5 Ω. If eight 10 Ω resistors are connected in parallel the resistance drops to 1.25Ω. Now 1cm$^3$ of silicon has a resistance of about 200kΩ (sand is silicon dioxide $SiO_2$). If billions of these sand particles are touching a Ground (Earth) rod, they are effectively billions of 200kΩ resistors in parallel and the Ground (Earth) resistance can easily reach less than 100Ω which is the requirement for Ground (Earth) rods of every home, office or factory.

Below is an EXCEL calculation of resistor in parallel following the equation:

$$\frac{1}{R_T} = \frac{1}{R_1} + \frac{1}{R_2} + \frac{1}{R_3} \cdots\cdots \quad (4)$$

| 1/200,000 | 1/200,000 | 1/200,000 | 1/200,000 | 1/200,000 | 1/200,000 | 1/200,000 |
|---|---|---|---|---|---|---|
| 0.000005 | 0.000005 | 0.000005 | 0.000005 | 0.000005 | 0.000005 | 0.000005 |

.........

This was done till column KQ in excel, meaning connecting 303 resistors of 200,000Ω each (average resistance of each grain of sand) in parallel, gives a total resistance of 662Ω. Thus, finer grain size means even more ‘resistors’ in parallel. And when this fine sand is further compacted, it provides even more resistors in parallel providing even lower ground resistance.

A wet ground will lower the resistance of each sand particle but it must be noted that mineral filled water conducts electricity but not distilled water. In this author’s experience one location which is even famous for flooding has a relatively high Ground (Earth) resistance because the water is from a river which has less minerals. Another location can be dry but is closer to the ocean so the ground resistance tends to be lower.

But why do electrons actually flow to the ground? Electrons like to flow to ground because the earth’s surface is mainly silicon and oxygen in the form of silicon dioxide ($SiO_2$). Both are insulators. Insulator atoms like to absorb electrons, which is why when a voltage is exerted on an insulator, there is no current flow because all the electrons in the insulator have been absorbed by the insulator atoms. So, there are no free electrons for the voltage to push. $SiO_2$ is a semiconductor which is more of an insulator so it is electron absorbing and able to conduct a flow of electrons a little. If the earth is a made of plastic it will not attract electrons even though the atoms within it are always searching for electrons, simply because a lightning for example cannot flow through it. A lightning strike onto it will form a molten crater of plastic instead. If the earth is a ball of iron which has excess electrons it will not attract electrons because electrons are not attracted to a heap of electrons on iron-earth's surface. So, because the earth's outer layers are mostly $SiO_2$, all flow of electrons, out of homes, factories and from lightnings are to the earth. In other words, the earth is a sink of electrons.

It is for this reason that almost all electrical appliances' bodies are connected to the ground with a green wire. The reason this is done is because in case a sharp portion of the body of the appliance somehow cuts the insulation of the live wire, the whole body will be live and a person touching it will get a shock. But because the body is connected to the ground, via an earth rod, the electrons will rather flow to the ground than through a human body to his or her legs and to the ground. The latter is a much higher resistance pathway to the ground and electrons always take the easiest path, therefore the person touching a live body of an electrical appliance will feel zero electron flow through his or her body. But note this insistence of all electrical appliance having a third grounding wire other than live and neutral is not followed in all countries. Most home gadgets like televisions have only live and neutral wires. The purpose of the green grounding wire which is joined to the body of electrical appliances is to ensure a human will not experience any shock touching an electricity leaking home appliance or electrical switchgear. This is shown in Fig. 21.

A 240V shock will be more serious than a 120V shock because the force (EMF) for the current to jump into the skin is double. A person dealing with electricity should not be allowed to wear any gold, silver or other conductive ornaments. There are even cases of 6V sending high current through a person's gold ring sometimes even taking the finger off. When this author graduated with an electrical engineering degree and went through the ceremony at South Dakota State University, USA, an oath ceremony had to be gone through where oaths like not designing anything which will endanger humans and down the list is one oath whose statement was, "I will never wear gold on my body."

The weakest point in the human body with regards to ability to withstand electric shock is the heart and brain. So, these two portions of the body should always be as far as possible from any electric conductor, especially while performing live electrical work. The CPR (Cardio Pulmonary Resuscitation) is performed on a victim of electric shock. The human body is basically an electric machine so when electron flow gets into the body from an external source, the body system is disrupted; the highest danger is that electrical signals from the brain that instruct the heart to pump is disrupted. The heart thereby 'forgets' how to pump. CPR is done to reteach the heart how to pump. If there is no blood circulation in a victim, the first thing to do is to shout for help to call the ambulance. Then, 30 pushes to a point one inch above the bottom of the sternum (meeting point of the ribs), two mouth-to-mouth blows while pinching the nose shut and then another 30 push. This cycle of 30 plus 2 must be repeated five times before performing a test of the victim's blood circulation. The recommended method of testing for blood circulation is by placing the pointing and middle fingers at the soft region of the victim's neck in-between the harder throat and the hard muscles at the side of the neck. If the victim still has no circulation, the process must be repeated till the ambulance arrives. If there is circulation, the victim must be placed in the recovery position; body lying on the side, bottom arm stretched out and other arm on this arm, top leg foot touching floor and over the bottom leg.

One of the main reasons electricity leaks to the body of a switchgear is when plastic insulated wires goes into the switchgear via sharp holes drilled into its metal body. Cable glands must be used whenever a cable is running through a metal hole.

Whenever a contract a job is to be done with the grid equipment, grounding must be done after isolation. This procedure is called Local Grounding (Local Earthing). There is a strict protocol to deal with grid equipment repair work.

1) Application and approval of work to be done.
2) On the morning of the work, a tool-box meeting must be held to ensure all has PPE (Personal Protective Equipment).
3) The Competent person (the person who has the license to do that particular electrical work) must call the Authorized Person (person licensed to switch off the appropriate circuit breakers). An added safety is for the Competent and Authorized person to repeat each other's phone statements.
4) A Volt Detector that looks like a pen whose tip lights up when it detects voltage is tied to an Operating Rod which is an electrically approved fiber rod which can be extended like a car aerial. This Operating Rod on which is tied the Volt Detector will be used to detect if the overhead line is really dead. The volt detector is also used to detect voltage on any grid equipment, a switchgear, a transformer, a reactor, a voltage transformer (VT) or a current transformer (CT) before repairing.
5) A ground rod is planted about 10m from the electric pole.

6) A flexible 35mm$^2$ grounding wire (specially designed for this purpose and must be approved by the electric company) is used to connect to the ground rod first then clipped on all the overhead lines. But it must be clipped to the N line first, then the other three L lines. including the street-lamp line.
7) Only then can the repair personnel climb up the steel pole with a fiber ladder (fully metal ladders are not allowed). And these personnel must have a full-body-safety-harness.

For (5) above, the ground rod must be planted 10m away because of two reasons namely Step Potential and Touch Potential. Step Potential means if the ground rod is planted just next to the metal tower, a lightning strike, say about three miles away on the overhead line will move to this Local Earth rod and if a person is standing next to the pole, there will be a gradient of potentials around the ground rod. If the person's leg is separated, one leg can be standing on one voltage level and the other at another voltage level and if this voltage is high enough, electron will flow through one leg and move to the second leg to ground; killing this person. But if the ground rod is planted 10m away, the gradients of voltage potentials on the ground near the pole where a person is standing will be too small to harm the person. Touch potential means since the pole is grounded and is at 0V. Say the local earth rod is planted a meter away from the pole, during a lightning strike, the personnel's legs will be standing at a very high potential ground and if he happened to touch the metal pole at that time, electrons will flow through his/her body and kill him/her.

It should also be noted that a safety procedure of walking in a substation is to ensure your legs are not separated too far apart. This is because if at the moment of walking in the substation, there is a lightning strike or a HV short circuit, there will be a gradient of voltage potential on the substation ground and if the legs are kept close, the voltage of the ground from one leg to another is too small to cause too much danger. Stones are filled as the top layer of substation to reduce conductivity.

# Chapter 3

## History of Electricity

In 1600 William Gilbert coined the word 'electric' which comes from the Greek word 'elektron' meaning amber. He wrote a book named, 'On Magnetism' which was published in 1600.

In 1744, a Dutch scientist named Pieter van Musschenbroek invented the Leyden jar and named it after his town of Leiden (Leyden). This was the first capacitor invented. He was basically trying to capture whatever is electric in a wine bottle. He accidently found that it stored more charge if he held it with his hand. This bottle stores static electricity and was used by early researchers to perform experiments in electricity. It should be noted that static electricity was known to mankind much earlier on. An Englishman invented a hand rotated machine to continuously rub materials to continuously produce static charges. The glass bottle was coated with a metal foil which doesn't reach the cork top. The jar is partially filled with impure water because pure water is not conductive. A metal wire passes through the cork top which closes the jar. This wire is connected to an external terminal which is a sphere to avoid losses by corona discharge. Early researchers believed the charge was stored in the water but Benjamin Franklin investigated it and concluded that the charge was stored in the glass container.

In 1752, the 46-year-old Benjamin Franklin (American) with the help of his son, flew a kite with a key tied to the bottom of a silk string. They also tied a thin metal wire from the key to a Leyden jar. He then attached a silk ribbon to the key and held it from inside a barn away from the rain. It was a rainy day and lightning struck the kite and electrons went down the wet silk string to the key and into the Leyden jar. Ben was unaffected by the electron surge because he was holding a dry portion of the silk string (insulator) in the barn. There was a U shape between the key and the point Ben was holding the silk string which kept the portion he was holding dry. This is critical in electrical engineering installations; water will follow along the string (or electric cables) from a long way up but will fall at the bottom of the U, thereby the portion of the cable at the next termination of the cable has little water. Thus, in all electrical installations, there must be a U just before a termination point of the wire.

When Ben later moved his hand near the iron key he received a shock because the electrons from the key went to ground (earth) via his body. An arc jumped from the key to his hand. The experiment proved that lightning was static electricity. Ben was lucky to have survived this experiment because a later scientist who replicated his experiment was killed by the lightning. From this experiment Ben came up with the idea of the lightning conductor which later saved his home from a direct lightning strike. One person managed to capture a picture of lightning striking the Eiffel Tower showing lightning choosing to strike only the copper lightning conductor at the highest point and not any of the metal works. The steel metal work is a conductor and is joined to ground via a pile. The pile has a mesh of steel (BRC) and goes deep into wet ground, thereby making it a good ground (earth). But lightning chose to strike only the lightning conductor because it can differentiate the resistance difference between the steel framework and the copper conductor which is the lightning rod. Today various shapes of lightning conductor tops have been devised to attract lightning better but the basic principle from Ben is that if a grounded (earthed) copper rod is placed at the highest point of a building, it will attract lightning because the lightning "brain" has only one thought which is to go to ground with the lowest resistance pathway it can find. If lightning can find a human body as a pathway, that will be chosen. This happened when three children played football on a rainy day in Kuching, Malaysia which lies on the equator and therefore has a high lightning rate. It was a big football field but the lightning chose to strike two of the boys who died instantly. The boys have blood in them which contains iron which is a conductor. The lightning can differentiate this resistance difference between four feet of air and a child's body and therefore chose the body as the easier path to enter ground (earth).

Ben made a mistake that electricity will move from the positive to the negative and we are still living with his mistake today. The direction of current he postulated is opposite of the actual electron flow which causes current. Books were expensive in those days so when later scientist discovered that Ben actually made a mistake they decided to leave it that way. When electrical calculations are made totally the wrong way around, the answer will still be the same. Ben coined many electrical terms like conductor, condenser (capacitor), battery, charge, positively, negatively, plus and minus.

In 1780 Luigi Galvani (Italian) put a frog's leg in-between two different metals and detected a spark of electricity. But he believed the electricity came from the frog's leg and called it, 'animal electricity'.

In 1800 Alessandro Volta (Italian) realized the frog's moist tissues could be replaced by cardboard soaked in salt water. He disagreed with the notion of Galvani of 'animal electricity'. He studied an earlier British scientist who cut out an electric eel and found there were repeating patterns inside. He was fiddling with two coins of different metals and placed them on his tongue and experienced what he knew to be electric spark. People of those days knew of electric spark and it was mainly used as an entrainment for the rich, rubbing ember and demonstrating the spark in a wine glass etc. All people who did this or even the intellectuals like Galvani and Volta were called electricians. From that feeling on his tongue he decided to get one metal, place a cloth soaked in salt water and then another metal and repeated this pattern, just as in the electric eel. When he connected electrodes at both ends he saw sparks but it was not ending as all sparks till that time in history. This device had the ability to create sparks continuously. He thought this is continuous like a river so he coined this flow to be current as in the flow of water in a river. The unit of current is ampere or 'amps' for short. The flow of electric current is observed when a voltage (a force) is put across the conductor. When this same voltage is exerted over an insulator no current results because there are no free electrons to carry the current in an insulator. One ampere is defined as $6.28 \times 10^{18}$ charges per second. Imagine that many germs or viruses entering the body, which is why one ampere entering a human is enough to kill him or her. When current flows in a conductor, heat is produced because of the resistance to the movement which is caused especially by the vibrating ions which are atoms with missing electrons. These vibrating ions causes the electrons to also vibrate. The degree at which electrons vibrate is proportional to the resistance (R). As more electron flow is required to run bigger machines, the cables must be of bigger size. In other words, the size of cables needs to be increased to cater for higher current flow.

Georges Leclanché (French) invented a more stable battery called Leclanche cell in 1866 which was widely used for telegraph; which was invented by Samuel Morse around 1835.

In 1784, Henri Coulomb (French) discovered an inverse relationship between the force between electric charges and the square of their distance. Actually, the full force between the charges is the product of the magnitude of the charge in each particle divided by the square of the distance between them.

In 1820, Hans Christian Ørsted (Danish) noticed a compass needle when placed near a current carrying wire. So, he is credited with the discovery of the connection between electricity and magnetism.

In 1820, André-Marie Ampère (French) showed that parallel wires carrying currents attract or repel each other, depending on whether currents are in the same (attraction) or in opposite directions (repulsion). This laid the foundation of electrodynamics.

Faraday (1821) was different from the above scientist in being from a poor family. He was mostly self-educated. At age 14 he started at a book making factory. Here he read all the books in print and became quite knowledgeable. The chief scientist of England at the time heard of his diligence in learning and took him in as a technician. Now he had probably one of the most advanced labs in the world in his hands and utilized his time performing experiments. He was not treated right especially by the wife of the chief scientist. He fought through this and eventually became better than the chief scientist and the queen of England wanted to give him a 'Sir' title but he refused. He is currently generally regarded as the father of electricity. In 1821, Faraday invented the first electric motor. He started work on his motor soon after the discovery of the connection between electricity and magnetism discovered by Ørsted. He connected the positive of a battery to a wire, this wire goes to a switch and then goes on to a nail hammered into a piece of wood. On this nail is hooked a copper wire such that it is free to move. The freely hooked copper wire goes down into a container of liquid mercury. At the edge of the container is an electrode to which the negative of the battery is connected. A cylindrical permanent magnet (PM) was placed in the center of the mercury bath. Upon switching on this circuit, the freely hooked wire moved continuously around the PM. This was the first motor known to science. For the next few decades all motors had this structure i.e. a vertical motor. Then in 1834 a German man in Russia, Moritz von Jacobi created the first motor having the horizontal structure, basically replacing the freely hooked copper wire suspended in mercury with bearings. But because electricity was expensive the advent of motors had to wait for Thomas Edison's first electric grid. A staff of Edison, Frank Julian Sprague in 1886 invented the first practical DC motor.

Going back to Faraday, after his invention of the motor he deduced that if electricity can produce magnetism, magnetism should be able to produce electricity. So, shortly after he invented the first motor he also invented the first generator. There are many children and some adults also asking why can't we get a generator to turn a motor and let that motor in-turn drive the generator rotor thus producing non-ending energy. This is not possible because as current is taken out of the stator coils of a generator to turn the motor, the stator coils become more magnetic. As it becomes more magnetic it becomes harder to turn the rotor of the generator. This will cause the motor to have to work even harder to overcome this magnetic force which will draw even more energy from the generator stator coils making it even more magnetic and even harder to turn and so forth. There are many who dwell on free energy theories, this author believes it may be possible simply because Tesla said so but the above is not the way.

Faraday also came up with the Faraday's law of induction which states that induced EMF is directly proportional to the rate of change of flux. In 1831 Faraday invented the first transformer. So, Faraday basically created the most important inventions of the modern electric world, the motor, generator and transformer. The unit of capacitance is farad, named after him. He coined anode, cathode, electrode, and ion. Faraday constant is the charge on a mole of electrons (about 96,485 coulombs).

In 1824, William Sturgeon (English) invented the first electromagnet by winding 18 turns of bare copper wire in a varnished iron horse shoe. His 200 grams electromagnet could lift nine pounds of iron. Because he used uninsulated copper wire the turns around the horse shoe has to be a single layer so he could not increase the number of turns.

In 1830, Joseph Henry (American) improved the electromagnet by using silk thread insulated horse shoe as well wires with silk thread insulation. He wound many layers of turns of copper wire around the horse shoe. He built an electromagnet that could carry 2,063 lbs. of weight. Such electromagnets were used in telegraph sounders.

In 1827 Georg Ohm (German) came up with the relationship between Voltage, Current and Resistance: V=IR.

In 1831, Joseph Henry (American) created one of the earliest ancestors of modern DC motor. Unit for electromagnetic inductance is Henry.

In 1833, Wilhelm Eduard Weber (German) together with Gauss developed the first electromagnetic telegraph. Weber in volt/second is the unit for magnetic flux.

In 1834, Moritz Hermann Jacobi (German-speaking Russian) improved on Faraday's vertical motor and Henry's similar horizontal 'rocker' which moved back and forth, to create the first horizontal motor that looked like today's motor. He used battery and his motor to power a boat.

In 1890, John Ambrose Fleming (English) coined the Fleming right hand rule shown in Fig. 3 which is also known as the generator rule and the Fleming left hand rule as shown in Fig. 4, which is also known as the motor rule. In 1905, he patented the 'Fleming Valve' which is the first diode which gave birth to modern electronics. Fleming also coined the term Power Factor (PF).

Thomas Edison (American) was the next great inventor. In fact, he is the greatest as far as electricity is concerned. It is due to him that we have a lighted up night time and can continue to do proper work in otherwise dark rooms in even at night. Electricity would still been a lab science if not for his combining business, politics and engineering to bring out electricity to everyone. He created the 'Invention Factory' at

Menlo Park whose function was to make money out of inventions. He gathered many researchers to work as a team to develop products. These researchers need not be educated but are often the best skilled in their field. This was a new concept because previously scientist generally worked alone. Electricity, telephony, motion picture and a host of other things were mostly inefficient lab developments till he made them truly useful. Edison was the Steve Jobs of those days. Edison was issued 1093 patents which is the most awarded to any person till today. His favorite was the phonograph or voice recorder, the predecessor to all our recording devices used today. Edison spent 40 years continuously improving it. In the year 2000 leaders of science gathered to pick the most important person for the last 1000 years and Edison was chosen as number one. His recommendation was just hard work, not to give up on the problem and common sense. He said genius is one percent inspiration and ninety-nine percent perspiration. He started a company called the Edison Electric Light Company with J.P. Morgan who was the richest person in USA at that time and Vanderbilt of the railroad fame. This company was started before he invented light bulbs. Today it is called General Electric (GE) one of the top companies of the world and the biggest in the electric industry. The next few in ranking are Siemens, ABB and Schneider. A typical electrician needs to know these companies because if they are consulted or their parts used, the likelihood for project success will be much higher. Companies like those have reputation and a bad part with their name on it somewhere in the world, reported by customer will affect their worldwide sales. On the other hand, unknown companies can even sell at much higher prices, making incredible specification claims but might be uncontactable or closed down by the time a serious complain reach them.

Edison's lab was the predecessor to current labs such as the Google lab, Apple lap, Microsoft lab and the Facebook lab, where he stated that there are no rules as long as inventions can be accomplished. And, he created this lab way before people had an inkling of these ideas. Edison's last patent before he died was artificial rubber. Edison was still experimenting till the last moments of his life.

All the scientists prior to Edison worked mostly in labs. Edison went to school for only three months starting at age six. Here he would always ask, 'Why?' to his teacher. The teacher got so annoyed that one day that he scolded him for interrupting the class. He went home and told his mother this. His mother took him out of the school after a subsequent quarrel with the teacher and decided to teach him herself at home. Despite this, Edison cannot be considered uneducated because he was an avid reader. He got this habit from his parents, who were avid readers themselves, and often read book aloud to their children. Edison's parents bought a science experiment book and Edison performed all experiments mentioned in it. Thenceforth, he was more into experimentation than reading and built a lab in his home for this.

At the age of 12, Edison started work at a train station. He used all the money earned to purchase equipment and chemicals for his lab. One day Edison jumped to push the train station manager's son away from an oncoming train. For this valiant act, the station official rewarded him by teaching him how to operate the telegraph which was invented nine years before Edison was born by Samuel Morse. He was also offered a train cabin. He moved all the chemicals from his home lab to this train cabin.

He later started a newspaper from this cabin. His newspaper reported events of many towns which the train transverses while competing newspapers only had single-town news. It is because of this that his newspaper business was quite successful. It seems like a simple victory but in those days, it was very innovative to realize that trains had suddenly brought in a new opportunity and he took full advantage of it. In today's age it is equivalent to finding a niche in the fast expansion of connectivity worldwide.

With this money he made in the newspaper business, he procured much more chemicals to do scientific experiments since chemicals were the main science of those days. One day he jumped onto a train and the station manager caught him by his ear. He said he felt something snapped in his head at that moment and he lost much of his hearing ability. Much later when he became a rich man, doctors said his hearing problem can be cured but he refused treatment stating that deafness helped him to concentrate on his work. The deafness drove him to reading; he would go into a library and read all the books in them. His train career ended the day he caused an explosion in his cabin lab causing the station manager to throw him off the train.

Edison's next job was being a telegraph operator in Canada. His job was to send hourly signals to Toronto. So, he built a device that would do it automatically. This was his first invention. Later he invented an electric vote recorder and displayed it at a Science Fair in New York (NY). People at the NY exhibition were impressed but no body bought his invention. He was frustrated and made a promise to himself that from now on, anything he invents must be sellable. Thus, he is today known as the 'First Engineer'. A scientist can invest time and money on abstract science but an engineer must make money, thereby an engineer is half scientist and half businessman. Such a philosophy of Edison is what drives all high-tech companies of today. Engineers in those companies must project earnings for every research or investment they do.

Edison strung a grid of underground wires in New York, making it the first grid in the world to supply electricity to power water heaters around NY. In those days fireplaces was the standard method of keeping homes warm in winters. But in NY, buildings were growing ever taller. If one resident had a fireplace mishap and burnt his home, hundreds could die. So, this problem was brought up to the Invention Factory of Edison which came up with an electricity powered water heater at the bottom of the buildings and pipes going to each home in the tall building. In each home, there will be a spiral of pipes providing heat to replace the fireplace. This was an optimum solution and was quickly mandated to be compulsory resulting in a quick expansion of Edison's business. Only later did Edison invent the Edison bulb to replace the other fire hazard, the kerosene lamp. Edison and his team tried out about 6000 materials before settling on carbonized bamboo as the filament in the Edison bulbs in 1882. In 1904, tungsten was found to be a better material for the filament. Edison would spend hours and days in his laboratory concentrating on some experiment or problem; milk, bread or tea was pushed into the room from under the closed door, but they were untouched until he solved the riddle that was in his mind. So great is the riddle that science demands.

Edison's business became so big that it spread all over USA and moved into Europe; this was rare in those days, when it was usually European inventions that moved to USA. Recently satellites have recently taken numerous pictures of the earth at night and these pictures were combined to get a view of earth without cloud cover. The whole earth can be seen to be lit up; of course, the advanced countries and cities are lit up brighter. And all this is due to Edison

In Edison's office in France a brilliant man named Nikola Tesla joined his company. The manager in France was so impressed by Tesla that he wrote a letter to Edison stating that he knows only two great men Edison being one and this man named Nicola Tesla being the other. Edison replied instructing Tesla to be moved over to the headquarters in NY. Tesla moved to NY with just a few cents in his pocket. The first task Edison gave Tesla was to improve the gensets to a particular specification. Though Edison was a great inventor who had an elite team of people working for him, they could not achieve those specs. It is because of this that Edison was so confident that Tesla cannot achieve those specs, and told him he will be given $50,000 ($500,000,000 in today's money) if he can achieve it. Tesla achieved it in a short while and went to Edison to collect the money. But Edison said it was just an American joke. The frustrated Tesla resigned. He found employment digging drains in New York. Tesla would inform his supervisor about his theories of science. This supervisor informed another rich man named Westinghouse. Westinghouse offered Tesla a lab and eventually $1,000,000 for most of his patents, including the AC system and the induction motor patent. The induction motor was invented by Tesla much earlier on. In those days carbon brushes of DC motors was of low quality but today a carbon brush in a DeWalt or Hitachi drill can last more than 20 years. So, the commutator system consisting of a pair of carbon brushes and copper electrodes separated by air gaps was the biggest worry for industrialist. A motor suddenly not running in a factory can cause lots of damage to products. Therefore, Westinghouse was confident Tesla's brushless induction motor will be easily sold. From the Westinghouse lab, most of the AC system we use today was invented by Tesla. Tesla lost a huge financial windfall because he tore up the contract which states that he would get a certain amount of money for every kW of electricity sold, which would have made him one of the richest person in the world.

Of course, Edison was not at all happy with the brewing AC system. DC was a business-friendly solution and actually a stable one. DC cannot travel far because it cannot be stepped up with a transformer.

How AC and transformer helped in enabling long distance electric transmission is as follows. Say in a hydroelectric power station, the power (P) is from the falling water. This power has two components voltage (V) and current (I):

$$P=VI \quad (5)$$

A hydroelectric generator produces electric power normally at about 11kV. A transformer is used to push up the V to a very high level, like 500000V (500kV). Looking at equation (5), this will cause the I to go down because P cannot change since it is the power of the falling water. The electrical energy then goes to the overhead lines where energy loss in transmission is defined as:

$$P_{loss} = I^2R \ (6)$$

Therefore, as power is sent to a faraway city say 1000 miles away; putting the low I into equation (6) will give a low power loss. Without this being taken cared of, the overhead lines will be red hot and by the time it reaches the city, there will be very little power left. Actually equation (5) is the same as P=VI but it is an equation generally used by electricians to calculate power loss because it emphasizes the importance of current and not voltage or power that plays the major role in power loss. Power loss in a single phase is given by equation (6) but in three phase the equation is as follows. In Fig. 1 below we assume Y connection but using Δ connection as in Fig. 2, the final equation is the same. $V_L$ is the line voltage $I_L$ is line current and $V_P$ and $I_P$ are the phase current. In homes in U.S. $V_P$=120V and $V_L$=208. In Britain $V_P$=240 and $V_L$=415V

$$P = \sqrt{3}\,V_L I_L \quad (7)$$

$$\text{But } V_L = \sqrt{3}V_P \quad (8)$$

$$P = \sqrt{3}(\sqrt{3}\,V_P)I_L \quad (9)$$

$$\text{So } P = 3V_P I_L \quad (10)$$

But $V_P = I_L R$ (11)

So $P = 3(I_P R) I_L$ (12)

But $I_P = I_L$ using Y connection

$P = 3(I_L R) I_L$ (13)

$\boldsymbol{P = 3I_L^2 R}$ (14)

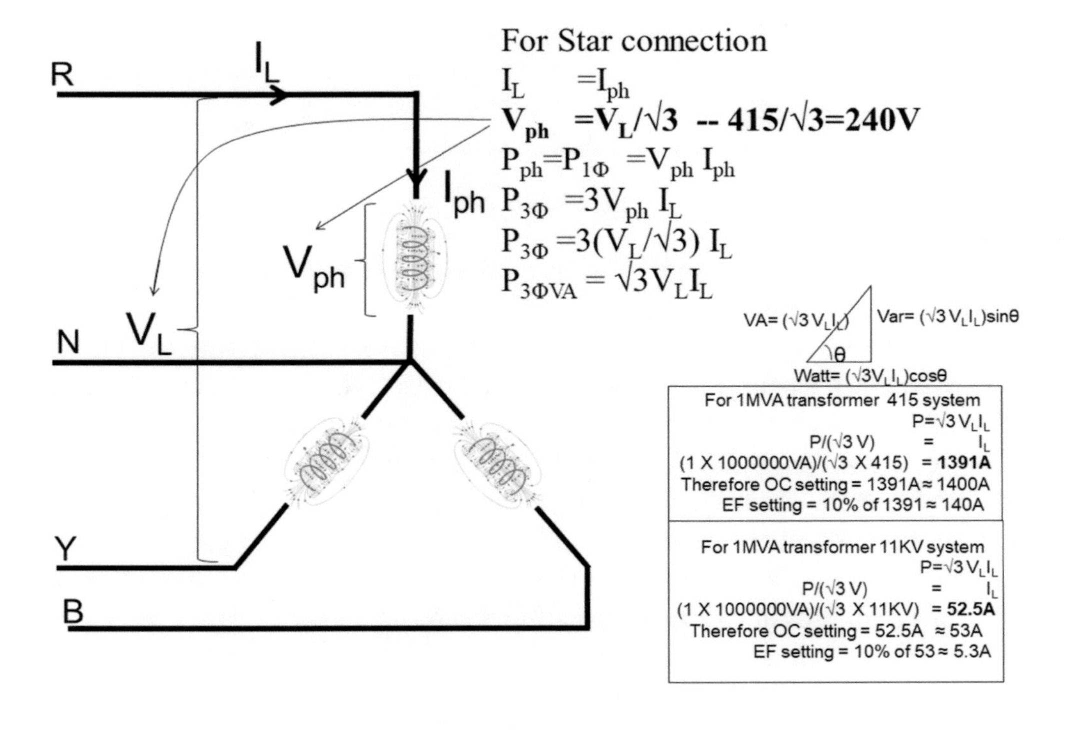


Fig. 1: Y connection

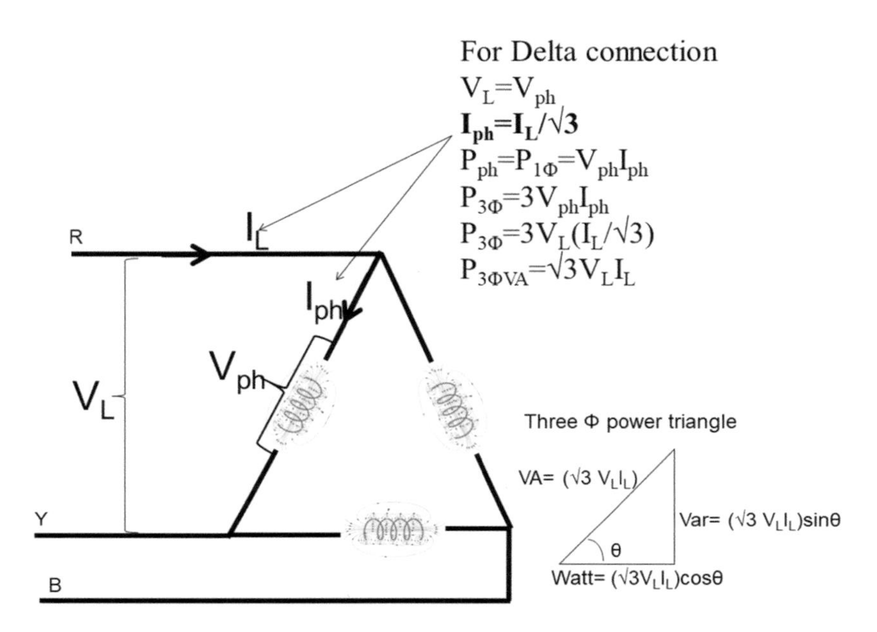


Fig 2: Delta connection

But Edison's business model was good; a DC genset could supply electricity only around a one-mile radius. He could sell electricity and gensets which made him lots of money. With the advent of Tesla's AC, a single genset can power the whole New York state. However, DC has one advantage in that it is very stable. If a DC line is running across a city, anyone can tap from it or send power into it as long as their generator is producing the same voltage. AC is comparatively very 'nervous'; voltage as well as frequency must be the same before joining a genset to an overhead line. That means a genset need to be synchronized to the overhead line current before joining. Also, lots of factors can cause phase shifts of voltage and current in AC which means the voltage and current do not travel at the same moment and these causes problems for which reactors (big inductors), capacitors banks, playing around with excitation of synchronous motor and many other methods are required to solve. So, Edison was right in stating about AC, that it is, "a useless complication". But he was most probably more motivated to say those words to defend his DC system because he can foresee a destruction of the DC system if AC is fully implemented. He was not thinking of the benefits to mankind as a whole. He knew DC had one big defect which is that the voltage cannot be stepped up easily with a machine as a transformer does for AC. Stepping up the voltage reduces the current enabling long distance transmission. AC is definitely an increase in complexity but the advantages far outstrip the disadvantages. Prime among the advantages of AC is the cost reduction in the electrical system and therefore the final bill people paid.

Edison used lots of resources to prevent the advent of AC, just as any businessmen would to secure their business. At least in the opinion of this author this is a misguided route to business and life; only activities that help other humans are worth pursuing. He put up newspaper articles stating that AC is dangerous and even built the Electric Chair to execute people using AC; just to show how dangerous AC is. But of course, a DC powered electric chair would also kill a person actually faster because DC is always at a high voltage while AC voltage goes high down to zero and goes high in the negative direction and repeats (sinusoidal). For a 50Hz system for example, the voltage will be zero, three times every 20 milliseconds. This is why it is easier to build an AC switch than a DC one. A high voltage AC breaker will wait for the voltage to cross zero volts before throwing the switch ON or OFF. The DC high voltage breaker was only invented by ABB in 2015. So, even with high voltage DC (HVDC), till 2015, the supply must be converted to AC before a breaker can switch it.

Despite the attack from Edison, Tesla got his break with the, "Chicago World's Fair". The USA government wanted to celebrate the 400$^{th}$ anniversary of Columbus' voyage to America and wanted to have a World Fair at Chicago. Quotations were given by General Electric (Edison and JP Morgan's company) as well as Westinghouse (where Tesla worked). But Westinghouse's quotation, using AC was about one quarter the price, so Westinghouse was chosen. Edison who got anxious about this, stated that his Edison bulb cannot be used for that project, so Tesla invented a double stopper bulb which was manufactured for only one year because the patent on Edison bulb expired the next year. At the exhibition, Tesla also displayed the early form of the florescent tube.

It was at the Chicago World's Fair, that humans from all over the world saw for the first time so much brightness in the night sky and this shot Tesla to fame. JP Morgan thenceforth decided to get General Electric to use AC with Tesla's help. JP Morgan's argument is that people in Germany were using Tesla's system to transmit electricity over 160 miles which was not yet achieved in the USA, despite, as Morgan said, "having Tesla right below their noses". Edison having lost the 'war of currents' went away from the electric industry and focused on inventing the whole movie industry and numerous other projects. His inventions were critical for the USA military effort in WW1.

Tesla was later free and was able to move to Colorado to solve the mystery in his mind. While working for Edison, he observed purple light emanating from electric overhead lines for a short moment after the big DC switch was turned on. After a while, the purple light would vanish. But if a man stood below the overhead line while the purple light was observed, it would travel down through the man to ground, killing him instantly. He could estimate that this purple light was not caused by electrons because it seemed to travel much faster than electrons. As mentioned earlier in this book, electron flow in DC is as slow as putty flow. Tesla used all his resources to discover what this purple light is and finally concluded it was energy emanating from ether (empty space); he called it 'radiant energy.' People always thought empty space is empty but Tesla stated that it is full of energy. This purple light emanates when a powerful DC is switched on. So, he created a Tesla coil using capacitors and inductors and a spark gap to achieve a continuous switching of a powerful DC current which replaced the big DC switch used at Edison's company, to achieve free energy.

Tesla then moved back to New York on the invitation of JP Morgan who heard about his achievements in Colorado. JP Morgan financed the big Tesla coil named Wardenclyffe Tower for $150,000. JP Morgan was only interested in having a way to communicate with Europe. Tesla already invented the long-distance radio transmission. In Europe Marconi is named the inventor but the US government have officially named Tesla as the inventor; Marconi used 17 of Tesla's patents to recreate it. He was somehow not caught for copyright violation by stating his system is in a sealed black box. Tesla told JP Morgan his tower can achieve telecommunication with Europe mostly with through-earth communications. But, before fully completing the tower, Marconi's radio invention came out and was basic and thus much cheaper. Also, at this time, JP Morgan went for a site visit to the Wardenclyffe Tower where Tesla informed him that his tower can be used for communication with Europe and also create free energy which can be broadcasted to the ionosphere. Ionosphere is a layer of ions surrounding the earth. Ions are atoms with more or less electrons than protons. If the number of protons and electrons are the same, it is an atom. In the ionosphere the ions mostly have more electrons than protons because these are gas or insulator atoms. Note all ions, even insulator ions are conductors. His plan was to send electricity to this conductive ionosphere where it will flow continuously. This electricity can then be tapped from the ionosphere for example by a home in Europe or a ship off Japan by utilizing a small resonating Tesla coil. JP Morgan asked Tesla how would his company charge (meter) the home in Europe or the ship off Japan for the electricity created by them. Tesla said it would be free, and this got JP Morgan very angry; businessmen do not want to give away things free. Besides JP Morgan was the main owner of most of the cable making factories in USA and wireless power will cause the closing down of all those factories. He immediately pulled out all funding. Actually, what Tesla was building was to enable not just electrical power and communication, he also stated that images or movies can be transmitted in real time across continents. Thus, Tesla was about to bring the world to beyond today's technology in the 1902.

Because of this, Tesla was dejected and used his little resources to build a car powered by this free energy. Others who saw the car were amazed that there was no battery or combustion engine, so they asked if the car worked on 'Black Magic.' So, now the car is called, "Black Magic car". Tesla later wrote how he made this free energy in a paper and cut it into a jig saw puzzle. He put a piece of the jigsaw puzzle plus, a cover letter into an envelope and sent it to all the major countries of those days. In the cover letter, he wrote that if you all become friends you will have my energy. This was his one little attempt to solve his frustration with the numerous wars among countries

Tesla also invented other things like free fertilizer. He said 78% of air is nitrogen, and fertilizer is just a way to get nitrogen into plants so he questioned why can't nitrogen be tapped from the air and used as fertilizer. He thus invented a free energy machine into which farmers have to hoe in soil and out comes nitrogen filled soil.

Tesla's induction motor can be made to last for a very long time with almost no maintenance. One can observe in industry today that the good quality ones almost never fail. This author has seen a 25 HP (1 HP or horsepower = 746 Watts) motor, made by Baldor, carrying 500 kg weight and running 24 hours a day, every day for 21 years. These motors get about an hour break each day because humans cannot load the machine fast enough. And the process requires spraying water (like a heavy rain) for 24hours a day and the motor is at the bottom-most point of that machine. This was at the Western Digital hard disk factory in Kuching, Malaysia. But it must be noted this author has also seen induction motors which look exactly like the Baldor motor emitting varnish gasses upon first energizing with 415V. Therefore, not all induction motors are the same though they look alike on the outside. Brand makes a whole lot of difference. It is mostly the difference in the quality of the varnish around the coils and bearings which makes the difference. A good electrical engineer or technician need to continuously keep up (i.e. coffee break talks etc.) with the latest information of which company produces quality parts, especially in the electrical space, but also including other hardware also. This is because projects can fail because of a deficient motor.

Imagine a Tesla world, one can drive as far as one can and will never need to refuel. The engine is an induction motor which almost never fails as described above. If car driver stops for a meal, the price of the meal is minimal because fertilizer is free. There will be no need for the electric grid wires because all electric transmission will be wireless. Even internal home wiring will be redundant because there can be a hub in each home sending wireless power to all appliance even heart pacers within a human body. This would be a utopian world with no greenhouse gasses; mother nature will love humans. But the huge oil and gas industry will collapse, electric utilities will collapse, the whole fertilizer business will collapse, cable factories will collapse and electric towers factories will collapse. This will give a hard time for the rich people of the world who make most of their money using these industries. The remaining industry will be the medical industry, but with the release of suppressed medical technologies, this earth will truly be heaven. Beyond this, only the behavior of people can damage countries and then the earth as proven in all super wealthy countries where various illicit drug and other additions are common for the rich. If Tesla's systems were adopted, no one will be too rich and it will take very little money for the poorest to survive. So, everyone will be about equal in wealth. Of course, capitalist will not agree to implement Tesla's innovations; they want to keep their status quo of being above the rest in wealth and power. So, Tesla's name was marred as far as possible and even taken out of science books for a time.

Today researchers all over the world are peering into Tesla's patents hoping to replicate his wonders. In the 1920s he achieved frequencies which the latest telecommunication companies cannot achieve today. After the decommissioning of the Wardenclyffe Tower by JP Morgan, Tesla himself stated, his inventions were ahead of his time but one day humans will use his inventions; humans needed to see the consequences of inefficient and earth polluting methods before finally adopting his methods. As far as the usage of his induction motors, which run our home fans, air conditioners, factory motors all the way to ship propellers and the largest cranes in the world; his prediction has already come true. The AC system he invented is currently the most expensive installation of mankind today; worth around six trillion dollars. Therefore, it is a safe prediction that all his inventions will one day be used by all of humanity.

It is possible the God sent Tesla to solve the problem of mankind at the right moment in history but the rich businessmen were not willing to forgo their wealth and power to achieve a pollution free world. An IEEE magazine reporter asked Albert Einstein how it felt to be the smartest person alive. Einstein replied that he does not know and stated that question have to be asked to Nikola Tesla. So, while most of humans assume Einstein is one of the smartest people who ever lived but Einstein himself knew that Tesla is smarter than him. Another interesting thing about Tesla was echoed by Steve Wozniak (of Apple) who said all the best things he did came from not having money and not having done it before. Some companies have huge research budgets but it often takes a 'Tesla' to get things done with often a miniscule of that budget.

# Chapter 4

# DC and AC currents

DC current can be equated to a pipe carrying water. The pump is equivalent to generator. If the pipe is soft plastic and is pressed, the water within will have difficulty flowing; this is equivalent to a resistor (or capacitor or inductor) which slows the flow of current within the wire.

AC current is actually moving back and forth. It is much more complicated to understand, which is why Edison called it, "a useless complication". But the world has adopted it because the benefits far outweigh the complication. In a generator, as shown in Fig. 8 (say a big generating power station), looking at only the red coils, as the rotor magnet passes the top clockwise coil, a positive high voltage is generated and as the rotor magnet passes the bottom anticlockwise coil a negative high is generated. This is how a sinusoidal waveform of AC is generated. The positive high and negative high actually means electrons moving back and forth in the outgoing cable. The current wave follows the voltage wave. You can imagine that the voltage is the force and electron movement are tiny balls being pushed by the force. At the generator, the tiny balls move exactly in proportion to the force being exerted upon it. But far away from the generator the balls move slower than the force exerted upon it. This is just like the balls are not following the instruction of the force exerted upon them perfectly or are lazy to follow the force strictly. This is called phase shifting among the V and I wave. Capacitors, Inductors and many other methods are used to bring the V and I wave back in phase or else the cables of the grid will heat up. To understand how this can happen, you have to look at equation (3). CosΘ value indicates the shifting of V and I. With lots of shifting, CosΘ value goes down. Since V is fixed by the power company and the P is fixed as the power a particular machine will draw (say 10kW) only I can increase to overcome a decrease in CosΘ. And I is the only factor that heats up cables. Basically, out of the important electrical factors, of P, V, I and R only I determines cable size. In effect this shifting of V and I wave is one of the biggest problems of the AC system and lots of investment are required to solve it on a grid.

So, theoretically an electron experiencing AC voltage is actually moving back and forth at a frequency set by the power supply company. Today there are two main frequencies used worldwide, which are 50Hz and 60Hz. 60Hz was the design of Tesla as he created the AC system. He turned a magnet past a coil of wire which was connected to a bulb. As he rotated the magnet pass the coil, the bulb lights up and switches off after passing the coil. He turned the magnet faster and the bulb switched on and off faster. He noticed that at 55 Hz the human eye cannot detect the flickering of the bulb. So, he decided that 60 Hz is the frequency to adapt. Currently, 60 Hz is used by USA, Canada, Japan and a few other countries. Japan, because they had a closed economy and when the emperor decided to open up, they surveyed the world to adopt systems from countries which were top of their fields. Since Tesla was in USA, they adopted the USA electrical system. Most of the rest of the world uses 50Hz because the Germans who always kept up with Tesla's innovations (Tesla was educated in German-speaking Austria) liked the metric system of 10, 100, 10000 etc. and were not so comfortable with 60 that Tesla decided upon. They decided that 100 divided by two was easier to remember. The British were the world power then and did want to adopt innovations created by a "ran-away-colony" country. But they respected the Germans enough to follow their system and Britain adopting it means, "The empire on which the sun never sets" adopted it and so did all other European countries' colonies. This 50 or 60 Hz is a multiple of the rotating speed of the generator rotor following equation (22). In USA and Japan, the generators turn at 60 cycles per second. Basically, steam turbines (ST) turn at 50 cycles per second just as motors in a factory 1000 km away will but for hydro and gas turbines turn at multiples of 50 Hz following equation (22).

As long as AC is flowing in a cable, a particular electron will stay at one point of the cable. Only if the AC supply is switched off will the electron be free to flow through the wire according to various external pressures such as heat, electromagnetic fields etc. This back and forth movement of an electron causes the next electron to move back and forth; continuing along the whole grid. This back and forth movement of electrons is utilized to run motors, bulbs, heaters or any other electrical load.

Fig. 8 shows the coils in a generator. Induction motor which was invented by Tesla, utilizes 68% electricity generation for industrial use today. A three-phase induction motors has the same coiling structure, within its stator as a generator but does not have a magnet at the rotor. So, as the generator passes the red clockwise coil, the red clockwise coil in the induction motor say 1000km away becomes a magnet and when the magnet passes the next coil in the generator, the next coil in the 1000km away becomes a magnet. The speed of current reaching the 1000km away motor coil is about 0.003 seconds ($1000^3$m divided by $3 \text{ X } 10^8$, which is the light speed – electric current flow almost at the speed of light). Therefore, in a three-phase induction motor, three phase wires sequentially magnetize the six coils (because most induction motors are four-pole motors) located in the stator of the motor resulting in the turning of the squirrel cage rotor. In a four-pole motor, each phase wire is turned into a clockwise coil and then to an anticlockwise coil. In some cases, the motors are Star connected and in other cases it is Delta connected.

In Star connected induction motors one phase wire (R, Y, B or L1, L2, L3 or R, S, T or etc. in different countries) is turned into a clockwise coil then into an anticlockwise coil, then the wire goes to a three wire Star point, with the other two wires coming out of the other two sets of two coils. At the Star point, the voltage is zero volts because the voltage level of the three phases at one moment of time can be positive or negative and adding the three voltages at any one moment always gives a zero value. Therefore, in a Star connected motor, the voltage across each coil is phase voltage (120V, 240V, 220V, 110V etc. - slightly different in different countries). This is because the coils are connected to phase voltage at one end, the current then flows through the clockwise coil then the anticlockwise coil and then to the Star point (zero volts). This is called four poles because at each end of one coil is a N and a S. So, there are four poles (N, S, N, S). Tesla demonstrated this sequential magnetization of solenoids in a circle to get a rotor turning in his famous Columbus Egg demonstration at the Chicago World Fair. The egg rotor was made of copper and it rotated as the solenoids were energized (magnetized) sequentially around it.

If the induction motor coil is connected in a Delta, one phase voltage goes to the clockwise coil, it then goes to the anticlockwise coil and then meets the second phase wire. When two phases are connected to a load (one phase at the beginning of the first coil, the wire continues to the second coil and at the end of the second coil is connected the second phase), the voltage is $V_{phase}$ X $\sqrt{3}$ =208V in U.S and 415V in Britain. These later voltages are called the line voltages $V_{line}$. So, basically, all the coils in a Delta connected motor have Line voltage while in the Star connected induction motor the coils are energized with Phase voltage. So, with that extra force (V) the Delta connected motors have more power. But Star connected motors are still used by some because they do not want their motors to be that powerful.

# Chapter 5

# Generator Principle

The generator works on the basic Fleming Right Hand rule. As taught throughout school, the three fingers of the right hand spread at right angles to each other represent the Force (F), Magnetic field (B) and the current (I). This rule states that the thumb, the forefinger and the middle finger represents the F (Force – this author feels this is wrong and prefers to call this the varying **F**ield density), B (Magnetic field from magnetic North to South) and I (the direction of the current in the conductor). Remember this by holding the right hand as a gun and the three fingers are FBI (American FBI use guns) top to down. This is shown in Fig. 3. Fig. 4 is the opposite of this, the motor principle. When a magnet passes a conductor, a current is induced, following Fleming's Right Hand Rule as shown in Fig. 3. Consequently, when current is flowing in a wire in-between a N and a S magnet, there will be movement of the wire, this is the motor principle. In electrical engineering there are many two rules like this it, it is easier to remember one and know that the other is the opposite. This author for example just remembers that left is for motor.

If more current need to be generated, one way would be to have a long wire and a long magnet passing it; but this is not practical so, the long wire is formed into a coil and a short magnet and powerful magnet is passed across it to derive the high current required.

Electrons can be equated to tiny magnets so when there is a flow of electrons in a wire, there is an electric field around it. In 1905 Einstein wrote a paper stating that electric fields and magnetic fields are the same phenomenon viewed from a different plane. In a generator, a coil of wire is spun in the rotor and DC is sent to this coil via carbon brushes. This turns it into a magnet (solenoid). In smaller generators, a permanent magnet is used as the rotor. In big generators permanent magnets cannot be used because the heat within the motor is too high and a permanent magnet will turn back into iron (Fe) when it is heated. But the magnetic strength of permanent magnets is increasing over time, mainly due to innovations at Hitachi. So, today increasingly higher power output generators are using permanent magnet rotors. However, the only advantage of using permanent magnet rotors is that there is no need to send DC current to the rotor via carbon brushes, so as to turn it into a solenoid. But with the huge increase in carbon bush technology, where even the carbon brush of a good-brand drill (like Hitachi or DeWalt) can last for up to 20 years, the justification for a permanent magnet rotor has decreased. There is a chapter on magnetism later but briefly Fe is magnetic because there are tiny regions within Fe called domains where the free electrons flow in the same direction and therefore parallel to each other. Thereby each of these regions or domains are tiny magnets. In Fe, the orientation of these domain sized tiny magnets is different, so Fe is not magnetic. But when a permanent magnet is swiped over an iron needle for 40 times, the direction of the magnet in all the domain aligns turning the Fe into a magnet. In a solenoid, the electrons are moving parallel to each other in the coils which is similar to electrons spinning parallel to each other within each domain. In a permanent magnet all the domains' magnetism aligns. This is why a permanent magnet does the same attraction for Fe as a solenoid. The advantage of a solenoid within a generator is that the direction of electron flow is fixed and is much less affected by heat than a permanent magnet.

The free electrons electron spinning parallel to each other within each domains of Fe makes each domain a tiny magnet in one direction. Some are asking why isn't there one particular direction to which there is a slight majority of domain magnets orienting to, resulting in iron attracting another piece of iron a little. The answer to this is analogous to throwing a coin. If we throw 10 times we may not get five heads and five tails, but if you throw it 100 times, you will get closer to 50 heads and 50 tails but if you throw a billion times, the it will be quite surely 50% heads and 50% tails. So, with billions of domains, the likelihood of no one direction having a little more magnetism is surer. When a magnet is used to swipe over a piece of iron, what is happening is that most of the domains get to be oriented in the same direction causing the iron to be a magnet. This is the similar explanation to a LASER light. If you look directly at white light, one sine wave will be along the Y-axis (90 degrees) another sine wave will be at zero degrees another will be at 30 degrees and many different planes. The LASER device makes all these waves in multiple angles to be in one direction becoming a powerful LASER beam that can be made strong enough to cut a thick iron sheet. In a piece of iron, when all domains' magnetism is oriented to be in the same direction, the iron can be a very strong magnet.

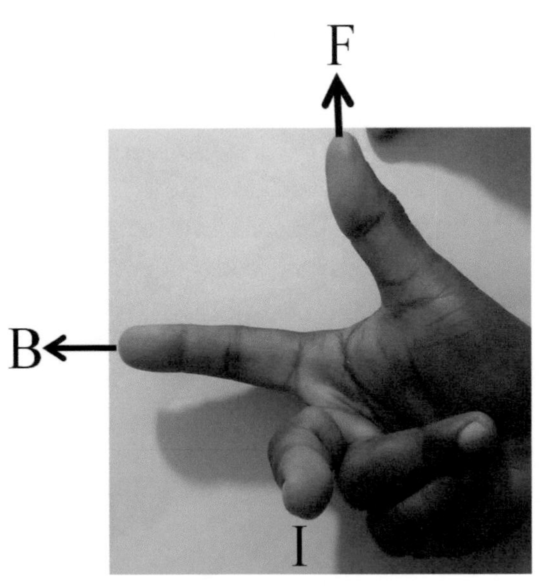


Fig. 3: Fleming's Right-Hand Rule

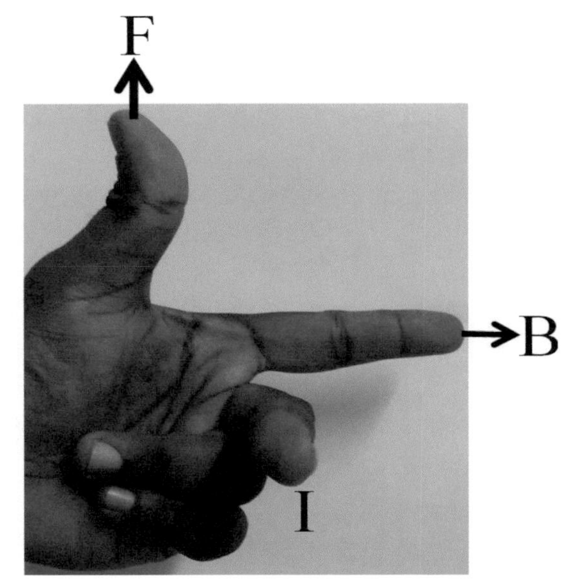


Fig. 4: Fleming's Left-Hand Rule

What is really happening when a magnet is passed over a wire? When the magnet is far away from the wire, the magnetic field lines that the wire sees are few in number. As the magnet is passed closer to the wire, the magnetic field lines proportionately increase till it gets to be maximum as the magnet passes over the wire and then reduced as it then moves away from the wire. So, what the wire is experiencing is a varying magnetic field density. This varying field density is a prerequisite for current generation in the wire. Using this concept, the transformer and skin effect can be explained. There is a transformer topic later but briefly the transformer has two coils, the Primary where current is injected and the Secondary where current comes out of it. If DC current is sent to the primary, there will be no current generated in the secondary. The reason is similar to the fact that even if the most powerful magnet in the world (currently at the CERN lab) is placed above a wire, there will be no current generated in that wire. That wire will experience a tremendous magnetic field density but there will be not current generation in the wire. That powerful magnet must move in order to generate current in the wire because when it move a little, the magnetic field density the wire experiences varies and that generates current flow in the wire. Similarly, even if a super huge DC current is sent to the primary of a transformer, there will not be any current generated in the secondary because DC like HVDC (High Voltage DC) can have a huge field density but it is not varying so there is no current generation in the secondary. But if AC is sent to the Primary there will be current generation in the secondary because the field density is varying. This is because in AC, the current goes from zero field density when the current wave is at the zero-current point of the Y-axis. As the current increases, the magnetic field density increases proportionately. The field density then peaks as the current wave peaks, then the field density reduces back to zero and increases in the negative direction and peaks and then reduce in density back to zero. As the field density increases in the positive direction the current generated in the secondary is in the positive direction and when the field density increases in the negative direction the current generated in the secondary is also in the negative direction. So, in a transformer, there is AC in and AC out but by varying the number of coils in the Primary and Secondary, current and voltage can be stepped up or down. So, basically in current generation, the varying magnetic field density is of prime importance.

Skin effect is also described later in this book but briefly, it is an observation that in AC that electrons flow is at the outer skin and do not use the whole copper cross sectional area. How this occurs is that in AC, electrons are moving back and forth. Electrons are equal to tiny magnets. As these tiny magnets start moving in one direction, it achieves maximum speed at the center and reduces speed till it stops at the end of this direction. It then goes back other direction to where it started initially. As the speed of electron flow increase and peaks the center of this back-and-forth direction, the magnetic field density variation also increase to a peak as the electron velocity is highest and then reduces to reach zero at the other end. Therefore, there is magnetic field density variation, which means current generation will happen. Thereby there is current generation in an AC carrying wire and following through with your fingers and the right-hand rule, will show that this current generation is outward of the center of the wire. This is an unwanted current generation and any unwanted flow of current (for human machinery to work) is termed eddy current. Imagine each electron to be a magnet where the N and S are oriented in multiple directions. Using Fleming's Right-Hand rule, this will cause current generation in a multiple direction but all out of the center of the wire. The overall effect is to cause a pipe like flow of electrons near the surface of the wire. The eddy current generated also pushes all the electrons generated by the big generator far away to the outer diameter of the wire. This moving of electrons only at the outer skin of the wire when AC is sent through it is termed skin effect. In overhead grid cables, only about 8mm thickness from the outer diameter is used for conductance of electron. So, at the center of overhead lines, steel strands are placed for mechanical strength. It should be noted that the frequency of all grid frequency is 50 Hz (Britain) or 60 Hz (USA) but in airplanes it is 400 Hz. The reason why they increased the speed of rotation in generators of airplanes is because at higher rotor speed there is higher variation in magnetic field density resulting in higher current generation. Thus, higher current generation can be achieved with a smaller size and lighter generator which are critical factors in airplanes.

One more factor is that magnetic field travels best in iron. For electron conductance (equivalent to current conductance), the best is silver (Ag), copper (Cu) and Gold (Au) but for magnetic field conductance, the best conductor is Iron (Fe). But scientist do not call this flow of magnetic field, conductance, they call it permeability. So, Fe has the best permeability and there is no number two because iron is so cheap and number two is too expensive to be even considered as a main magnetic material. Iron has the highest permeability among elements. In fact, permeability of air is 1 and so is the permeability of most elements like silver, copper, aluminum or gold but the permeability of iron can reach 100 and with some addition of other elements it can reach even reach 10,000. So, in generators, current carrying copper conductors are surrounded by Fe to increase the permeability of the magnetic fields around the copper wires. This author has tried taking the Fe core out of a contactor solenoid, and energizing it. A screwdriver is then held near the solenoid and there is attraction to the solenoid but only with very little strength. When the core is placed back into the solenoid, the pull on the screw driver increases very greatly. So, this proves the flow of magnetic field is very much greater in iron compared to air. But Fe also conducts electrons and we do not want electron flow in the Fe. So, to prevent eddy current flowing in the Fe, laminates of Fe are used instead of a solid Fe; in-between these laminates is insulating varnish. This way, the unwanted eddy current generated can only flow within one laminate. This eddy current generates heat and any flow of current, especially in a not-so-good conductor like Fe will build up heat. But another even bigger problem caused by random eddy current is that it can be so big and when it is in the opposite direction, it will superimpose on the rotor solenoid's magnetic field, weakening it and therefore reducing the current generation in the stator coils. In the stator also, the random eddy current in a solid iron core could nullify the generation of current in the stator. Lamination prevent all these problems.

In a standard generator today three wires going into it because if a single wire goes into it, only one AC waveform can be generated and the space within the generator is not fully utilized. When Tesla explained that he wants three sets of coils in the generator which is in effect getting three generators out of one generator, others suggested why not put 19 or 27 coils. Later scientist found out that Tesla's mind is always right, three is the best because of the factors below:

1. The phase currents will cancel out one another and will sum to zero in the case of a linear balanced loads like induction motors which currently consume 68% of energy generated by industries. Here all phases use the same power but being 120 degrees apart they cancel out each other resulting in zero voltage and current in the return path or neutral. Fig. 7 shows the voltage graph. The current graph will be in phase with the voltage graph if the power goes through a resistive load. The voltage and current graphs experience a phase shift only if it travels through a capacitor (current graph faster) or inductor (current graph slower).
   Power to an induction motor uses the same amperes in L1, L2, L3 (R, Y, B) phases so the motor vibration is less. The vibration principle can also be applied to three phase generator and wind turbines which tend to have three fins.
   Three phase motors are simple because they rotate synchronously with the magnet passing the R, Y B coils in a far-away generator (or synchronously with a small slip as in induction motor).
2. Three is the lowest number to exhibit all the above properties.

For factor (1), this summing makes it possible to eliminate or reduce the size of the neutral conductor. That is, if L1, L2, L3 (R, Y, B) carry the same current and voltage, the sum is zero and therefore at the N (neutral) wire V=0V and I=0A. Even if the load is not an induction motor or not a balanced-three-phase-load, the N current is reduced. For example, at one particular moment if L1(R)=6A, L2(Y)=-4A, L3(B)=3A. The sum of the three phase current is 6+(-4)+3=+5A, so the neutral current at that moment is N=-5A. Since the conventional direction of N is towards the substation, the minus sign indicates at that moment of time, the current is moving towards the home. Electrons are moving back and forth but the voltage at the power station is high because it generates electron flow and the N loves to absorb electrons because N is grounded to earth and the earth loves electrons. But the circuit from power station to N can be seen as a rope; as the generator pushes, N pulls and when the generator pulls, N lets go. All devices are placed along this wire are powered by this rope which is pushed and pulled within them. An analogy is pushing and letting go a spring onto a wall; if this back and forth moving spring is utilized to operate some device, that would be analogous to the AC system powering devices. This begs the question, can we continuously pull from a ground rod, thereby providing free energy? Apparently, Tesla did this in the Wardenclyffe Tower. There are descriptions of massive grounding below the tower and electrons oozing out of the dome top of the tower which he somehow directed to the conductive ionosphere so as to travel to other countries to power homes, factories and ships.

Another case where people are in a bit of confusion regarding the above is in the calibration of MSB (Main Switch Board). The purpose of a MSB is protection. For a home that uses below 52A the system is a single-phase Distribution Box (DB), between 52A to 100A a three-phase DB is the protection system. Above 100A a MSB is used. And for a country or region, a Substation is used; note Substation for protection as opposed to Power station for power generation. Going from DB, MSB to Substation the equipment gets more and more complicated. In substation the latest computer and fiber optics technology is utilized. For a MSB the main protection system is simplified in Fig. 5. This is the layout of the protection CT (Current Transformers – to measure current) in a MSB. And Fig. 6 is the Protection Scheme (like a protocol). To calibrate the MSB first the L and N leads of the Earth Fault (E/F) Relay is disconnected. Then 100%, 150% and 200% of power company's approved load is injected into the bars to test if the MSB will trip in the period specified in the relay's manual. The O/C relay will not give the trip signal to the ACB (Air Circuit Breaker) which is the main cut-off switch for a building, at 100% injection; it only trips beyond 105% loading. So if it trips at 100% current loading, the O/C relay must be rejected and replaced. At 150% loading, it will trip at a specified number of miliseconds, at 200% loading, it will trip at another specified number of miliseconds. If the tripping time is out by a little, the calibration personnel can use a tuner in the O/C to bring it back to the specified timing. If it is out by a lot, the calibrator need to reject and replace the O/C relay. This all seem logical. We switch off the power company's incoming and use current generated ourselves to trip the ACB and detect if the tripping timing is correct.

But for the E/F relay, we inject the same points of the current carrying bar with 10% injection and the E/F will send a trip signal to the ACB. Many ask why is it tripping at 10% while normal consumption current is 100% (or at least 70-80%). Fig. 6 is the full setup for performing calibration. In the partial daigram in Fig. 5, all three phase wires are looped with the N wire. From the three CTs, current goes out via three wires and enters the O/C relay. Then three wires goes out of the O/C and all are looped and joined to the wire from the N CT. So all four wires are loped. This should cause this point to be at 0A. Note for a balanced-three-phase load the three phases will add to zero current at one moment of time but if the loads are not balanced, say it is +2A remaining then the N will carry -2A; resulting in +2-2=0A at this point. That leaves zero current at the wire going out of the four wires to the E/F relay in normal cases. But if there is an earth fault, like a live wire touching the body of a refrigerator connected to L1 (R) phase, some of the current from L1 (R) phase will leak to Ground (Earth). So, now the N wire has less current than the combined incoming current of L1, L2 and L3 (R,Y,B) at that moment of time. If this leakage is 10% current of approved load (by power company), the E/F relay will send a trip signal to the ACB.

Note the current coming out of the CT is much reduced from the actual current but is exactly propotional to the large current flowing in the three-phase current carrying bars. Relays cannot be injected with such large currents which is why the CT is used to propotionally reduce the current before it goes into the relay; electronics cannot work with large currents; they will get fried. The protection system is operated by the propotionate low current but the actual large current causes the triggering and tripping. In all power systems, CT are used to propotionately reduce current and VT (Voltage Transformers) are used to propotionately reduce the voltage so electronics and computer technology can determine what actions or reaction to take.

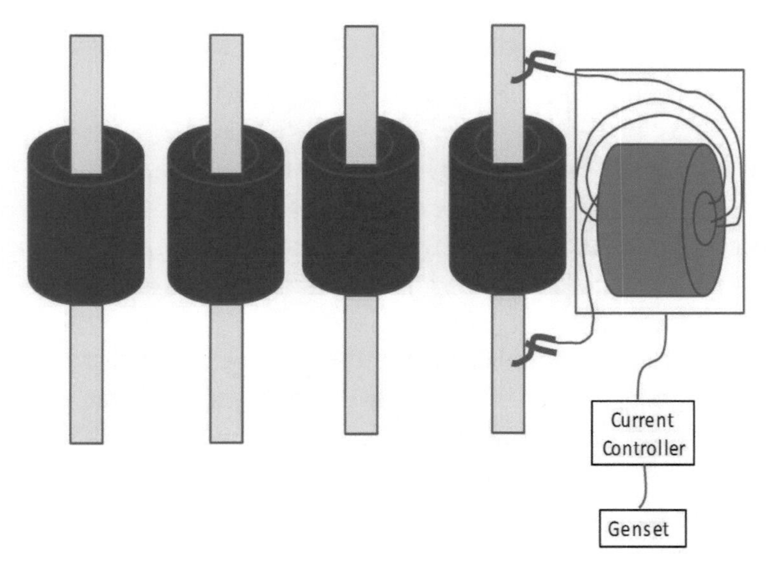


Fig. 5: Injection test where current generated by genset, current controller and big CT combination is injected into bars to simulate a fault condition

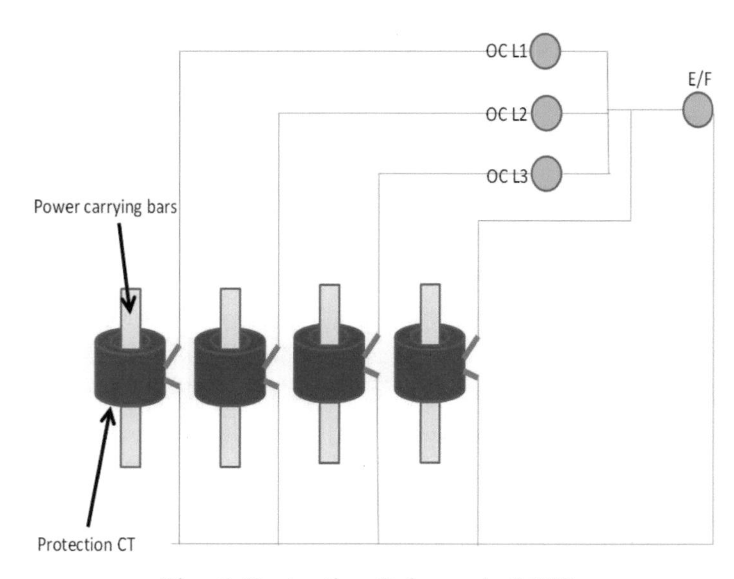


Fig. 6: Protection Scheme in MSB

The voltage is controlled to be at the fixed value by the power company. It is fixed to be 120V (240V British). In fact, that was the job of this author while he worked in a power company to maintain the main grid voltage to be 275 kV. This was done by continuously starting more generators or energizing or de-energizing reactors which are big coils located at the large substations (note coils=inductors=reactors).

If the highest voltage of the grid is maintained, all the voltages down the line will be maintained. Voltage is also controlled at LV (Low Voltage the opposite is HV or High Voltage) substations which are considered the distribution network, mostly by tap changes at distribution transformers. At 33kV substations the changing of the Secondary taps are done automatically (OLTC =On Load Tap Changer), in 11kV substations the changing of the taps is done manually. So, voltage at the consumer (home, office or factory) is fixed at 120V (240V British) per phase. This means at any moment of time, the sum of the voltages in the three phase wires is 0V. Note the three phases voltages sums to 0V at every moment of time but the three phases current at one moment of time do not add to 0A unless a balanced three phase load, like an induction motor is used. But for current, the sum of the three phase values plus N will sum to 0A unless a leakage has occurred. This can be observed in the Fig. 7; taking voltages to be full (top of sine wave) or half (half way down the sine wave). At 5ms L3(B)=+1, L1(R)=-1/2, L2(Y)=-1/2. So, at 5ms, $V = +1 - \frac{1}{2} - \frac{1}{2} = 0V$ . At 8 ms, L3(B)=+1/2, R=+1/2,

Y=-1. So, at 8ms, $V=+\frac{1}{2}+\frac{1}{2}-1=0V$ . Similarly. at 15ms, L1(R)=+1/2, L2(Y)=+1/2, L3(B)=-1/2. So, at 15 ms, $V=\frac{1}{2}+\frac{1}{2}-1=0V$ . Actually, at every moment of time the sum of the three phase voltages will be 0V.

Summarizing, for current in a in a factory where the only loads are induction motors, the current load on all three phases are always the same, therefore the N wire in that factory will record zero current. It will also read zero voltage because the voltages in the three phases will always sum to 0V plus N is connected to the Star point of the substation transformer and this star point is grounded with sometimes up to 40 round rods. In installations where equipment other than three-phase induction motors (balanced three-phase loads) are utilized, the current is not perfectly balanced and the remainder current will flow in the N wire.

By looking at the waves in Fig. 7, it is impossible for the sum of current in L1, L2, L3 (R, Y, B) to sum up to a value beyond the peak current in any of the L1, L2, L3 (R, Y, B) phases. Of course, Fig. 7 is the voltage waveform. The current waveform is in most cases controlled by the power company to be in phase (crosses the X-axis at the same times) with the voltage waveform (using PF compensation techniques) but the height of the wave (amplitude) of the wave of each phase is not fixed as in the voltage waveform because voltage is controlled by power company. The current waveform amplitude is controlled by the load. Say a load in a single-phase home connected to L1 (R) phase may have lots of high electricity usage equipment and another other single-phase home connected to L2 (Y) phase home may have very little electricity consumption. This will result in the current waveform for L 1(R) phase to have a high amplitude and the waveform for L2 (Y) to have a low amplitude. In the later part of this book there will be details of a new phenomenon called harmonics in electrical systems caused by many modern equipment which changes the fact that N current cannot be higher than the current in any of the three phase cables. But in most cases the N (return) need not be three times the size as is required for the return water pipe taking water from three incoming pipes.

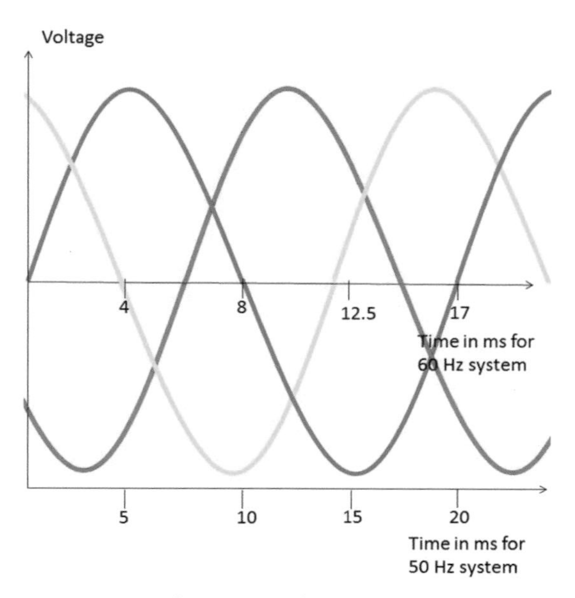


Fig. 7: AC sine waves

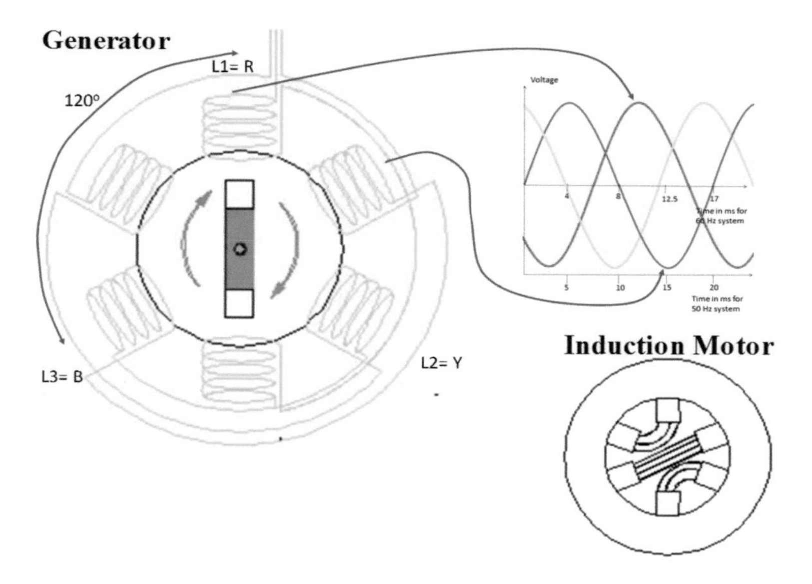


Fig. 8: Three phase generator and induction motor on right

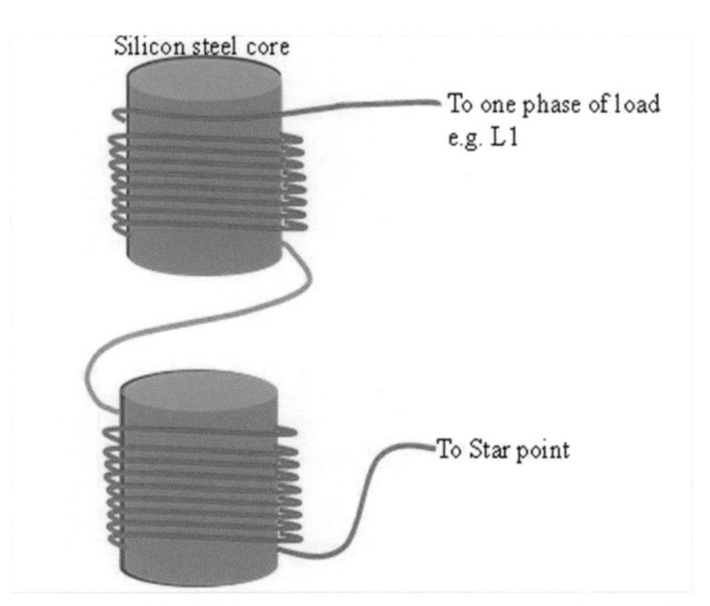


Fig. 9: Winging method for three phase generator and induction motor

The three wires are named L1, L2, L3 (R, Y, B) forms six coils; each wire forms a clockwise coil as well as an anticlockwise coil as shown in Fig. 8. To remember the polarity formed by the coil, it is looking at electron flow (opposite of current flow), a clockwise turn will form a N and an anticlockwise turn will form a S. One end of all three wires is tied in a knot called a star point. Note, as mentioned previously there are three ways to achieve zero volts:

1) Joining all three phases together
2) Neutral
3) Ground (Earth)

This star point is then connected to a transformer before going to Ground (Earth); thereby providing a high impedance Ground (Earth) or elevated N Ground (Earth), for purpose of protection. The other end of the L1, L2, L3 (R, Y, B) wires supplies power to homes and industries. This is depicted in Fig. 11. The output of big power generators are generally at 11kV to 13KV. Some are asking why 11-13kV when you need to immediately step it up to 275kV at the power station's substation, why can't we generate at 275kV? The simple reason is if a power station is operating at 275kV, it is simply too dangerous for the humans working in that power station.

Three Φ motor with eight pole per Φ
The N is connected as a Star point in all big generators

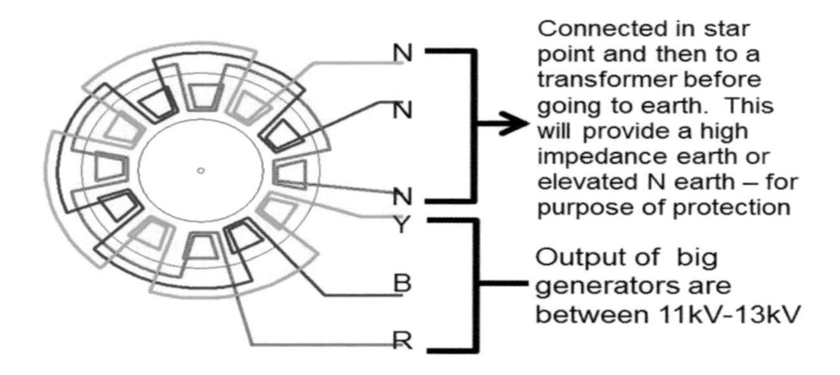


There is a CT on each of the six wires. Terminals of these CTs are connected to protection system

Fig. 10: Wiring in a generator

Fig. 11 depicts the generation of the current and voltage at the generator. Assume, only one wire goes into the stator (the stationary outside) of the generator, it forms a clockwise coil on the top and an anticlockwise coil at the bottom. As the rotor (the rotating inside) magnet passes the top coil, electricity is generated as a sigmoid curve. Initially a low voltage as the magnet approaches the clockwise stator coil, the electricity gradually increases till it reaches peak voltage $V_p$ as the magnet is fully under the coil. As the magnet moves away from the coil, the voltage decreases till it reaches zero volts.

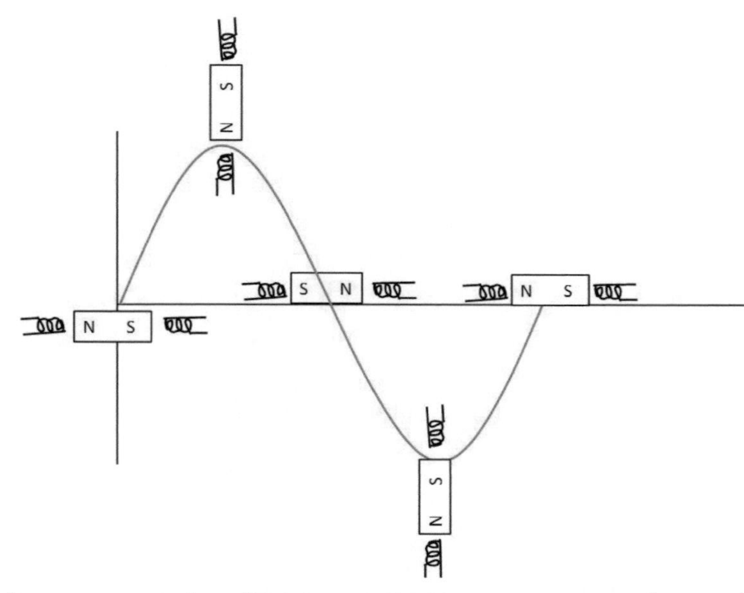


Fig. 11: Waveform of voltage generated as the rotor magnet passes one complete cycle around a generator

As shown in Fig. 8, when the magnetic N of the magnet passes the clockwise coil of L1 (R), a positive high is generated on the L1 (R) phase wire. The magnet then passes the L3(B) anticlockwise coil to generate negative high for L3 (B) phase wire. When the magnet passes the L2(Y) clockwise coil a positive high is generated in the L2 (Y) wire. Next it passes the anticlockwise L1 (R) coil generating a negative high in the L1 (R) wire, consequently a positive high in the L3 (B) wire and a negative high in the L2 (Y) wire. And goes back to produce a positive high in the L1 (R) wire. Thus, as the magnetic rotor physically turns a complete cycle, one complete cycle of AC waveform is generated. Many engineers who study just pure telecommunications read a lot about cycle but cannot imagine where the word cycle comes from. The cycle comes from one full rotation of a rotor in a generator.

Because the clockwise coil of L1 (R) phase and the next clockwise coil of L2 (Y) phase are 120 degrees apart, the phase to phase angle of the waveform is also 120 degrees apart. For example, the peak positive of L1 (R) to peak positive of L2 (Y) is 120 degrees apart. In a 50 hertz system used in most countries, the magnet turns 50 complete turns in one second or multiples of this using the equation (22). In USA and Japan, the generators turn at 60 cycles per second. Basically, steam turbines (ST) turn at 50 cycles per second just as in an induction motors in a factory 1000 km away will but for hydro and gas turbines turn at multiples of 50 Hz following equation (22).

As more DC current is sent into the rotor, it will generate higher current in the stator so the magnetism between the rotor and stator will be higher making the rotor harder to turn. Consequently less magnetism (less DC current) in the rotor will make it slightly easier to turn. In some hydroelectric power plants, say there are four penstock pipes coming down to four generators. The water is not allowed into one of the penstock pipe and the generator is run as a synchronous motor and playing around with the DC going into the rotor can cause a slight phase shifting in the entire grid. The way to remember this process is, more DC leads to leading Power Factor (PF). Leading PF means current wave ahead of voltage wave. This is because as more DC is sent to the rotor, it slows down and the rotating speed of the rotor is the proportional to the voltage wave. As the voltage wave slows down, it becomes in phase with the current wave because the almost all loads are inductive (motor coils) so the entire grids tend to be inductive and increasing DC in the rotor of the synchronous motor (generator run as a motor) will cause the voltage wave to be slower and eventually be in phase with the current wave.

The current wave is normally lagging because coils as is used in motors of industries, fans in homes or aircons in homes all cause lagging PF. Larger loads connected to the generator will also slow down the rotor because as more current is drawn from the stator coils of generators, they become more powerful solenoids whose magnetic field acts like strong strings to the magnetic field around rotor solenoids, thereby slowing down the rotor.

One complete cycle at 60 hertz takes 16.6666 milliseconds ≈17 ms (20ms British for 50 Hz) which is the period of the waveform generated. At low voltage, the single-phase peak voltage is controlled to be 169V (339V British). The RMS voltage or 120V (240V British) value is got by placing a voltmeter between any phase wire L1, L2, L3 (R, Y, B) and N or G (E) is derived from this by multiplying the peak voltage by 0.707 (this number can be remembered because there was a Boeing 707). Voltmeters can only detect the peak value. In analog voltmeters, the scale shown on multimeter is simply multiplied by 0.707 and in digital voltmeters, a simple multiplication by 0.707 is done to give the RMS voltage. The phase to phase voltage (placing voltmeter probe on any two of the three phases wires L1, L2, L3 (R, Y, B) is 208V (415V). The phase to phase voltage can also be mathematically derived by multiplying the phase voltage 120V (240V British) by the square root of three, that is $120\sqrt{3}$ =208V ($240\sqrt{3}$ = 415V British). The reason this is done is because the first use of electricity is for heating water. Edison's company did this with DC. Say two water heaters are placed is two containers holding equal volume of water. To one water heater is sent 120V DC. They found out that the heater supplied with AC must have a peak voltage of 169V to reach boiling point within the same time span. Technicians of those days did not want to learn a new number of 169V so Tesla said this is not a problem, just multiply 169 X 0.707 to get 120V. This is how the Root Mean Square voltage ($V_{RMS}$ = 120V) came about. Basically, RMS voltage has a sole purpose of making AC and DC have the same effective power with the same voltage numbers.

The three wires coming out of a big generator are at 11kV. It should be noted that the voltages in HV or overhead lines and generators are defined as the phase to phase voltage by convention not phase to neutral as is used in homes and factories. The 11kV out of big generators is stepped up to a very high value like 275KV or 500KV. The current is therefore reduced following the formula $P=VI\cos\Theta$. When V goes up, I has to go down because P is fixed as the power of the falling water in a hydroelectric power station for example. Neglecting $\cos\Theta$ which usually equals to one out of the power station, the formula reduces to equation (15):

$$P=VI \quad (15)$$

When V is increased with a transformer, I has to go down because the P is the fixed falling water power. The transformer therefore only plays around with the components of power which are V and I. And when this low current value is placed in the power loss equation (16),

$$P_{loss}=I^2R \quad (16)$$

results in a low $P_{loss}$ or power loss over the transmission lines which could span over 1000km before reaching loads in a city or other population centers. Here in the cities, it is stepped down to usable voltages. So, the whole purpose of a transformer is to reduce power loss over the lines. Since a transformer cannot work with DC due to the absence of a sinusoidal variation in current and thereby a variation of magnetic field density (FBI is needed to induce current in the Secondary), AC is chosen over the original DC system of Edison.

A typical high voltage (HV) tower has four cables carrying current per phase. This is called bundled cables, normally two cables bundled with spacers but it can be up to four cables bundled with spacers. With two cables bundled, it is four cables per phase (two on each side of the tower). That equates to 12 cables. If there are four cables bundled, that equates to 24 cables.

The current in each of these 12 wires is very low, ranging from 70A to 110A but it can carry the huge power used by the whole grid. This is because the voltage is very high, up to 500kV or 1000kV. 70-100A is actually quite low compared to a car battery which sends out 70 amps upon starting. But the car battery is 12V which is less than the 40V required to enter into humans. Note this author has tried touching the spark-plug wires conductor when the car is running and it was too painful to try again. Initially this seem illogical because the battery only supplies only 12V but there is an ignition system in a car to create an oscillating AC wave using and LC circuit, a constantly switching mechanical switch and a step-up transformer to produce 25,000-30,000V for the spark plug. Such is the voltage required to enable electrons to jump across a gap such as seen in a spark plug of a car.

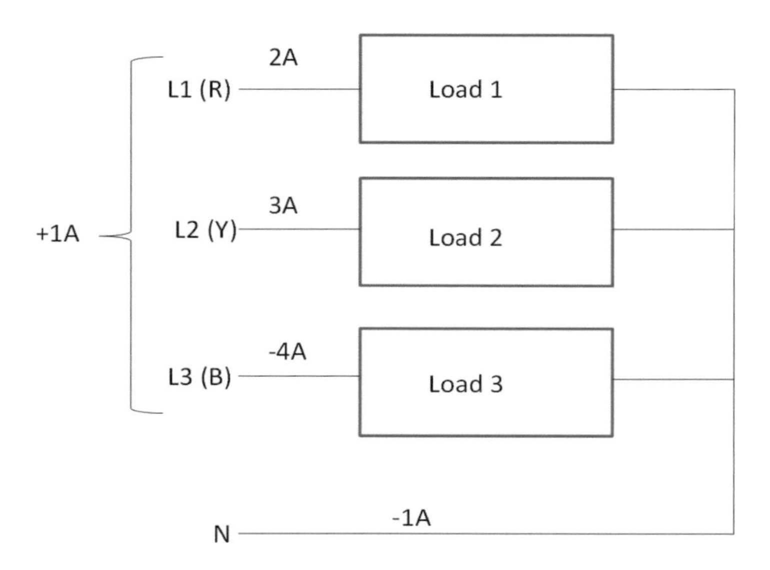


Fig. 12: Housing three phase wiring

As in Fig. 12, the power coming out of a generator is sent to homes where the L1, L2, L3 (R, Y, B) is connected to three homes. That is one phase for each home; this is an example, in actual case, sometimes two, three or four homes can share a single phase depending on the incoming wire size. Coming out of each home is N and all three N wires are joined such that there is only one N wire goes back to the Low Voltage (LV) substation transformer star point which is joined to ground (G (E)). If the N is broken in-between L2 and L3 phase (Y and B phase), L1 (R) phase and L2 (Y) phase will be shorted providing a voltage of 120√3 = 208V (240√3 or 415V British) for the first two homes. The N wire in these two homes has become live wires. Some new electricians are a bit confused about this because they assume with that color code (white wire in U.S. and black wire in Britain), the wire should be at zero volts but it is not the case if the case above happened. This will cause equipment in L1 (R) and Y (L2) phase homes to be damaged because they will be getting twice their rated voltage. Actually, it will not be exactly double, initially; each home will have parallel circuit (all homes and all installations are connected in parallel) but the two parallel connected homes will now be connected in series. So 415V supply will drop over each 'resistor' (home). If the two homes have exactly equal load, each home will get half the voltage and nothing burns. But if one home has a much higher resistance than the other, in a split moment of time, the high resistance home will get a higher voltage, blowing up all the equipment that home. Now there will be no more resistance in this home so the other home will get 415V powering all the equipment, thereby destroying all equipment in this home also.

If N is broken in-between L1 and L2 phase (R and Y phase), the L1 (R) phase home will not get any power since there is no return path to the substation transformer neutral. L2 and L3 (Y and B) homes will not know something happened as everything will be running as normal. L1 home people will then call the power company and nothing gets blown.

If N is broken in-between L3 (B) home and outgoing N, the N join of L1, L2, L3 (R, Y, B) homes are now connected as a Star point on the N side which is one of the three ways to achieve 0V. Note there are three ways to achieve 0V, namely, N, E and Star Point. So, when a Star point is forced upon the three homes it looks alright because in three-phase induction motors, the outgoing wires are sometimes connected in a star-point. But there is a problem in that in an induction motor the three coils use the same amperes while the three homes do not use the same amperes. As in Fig. 12, in the incoming current is (+2A) + (+3A) + (-4A) = -1A. At the star point, the current will be +1A so on the L3 phase there will be a (-4A) – (+1A) = +3A across the L3 home which seems alright but if the current is -9A at star point and +9A at L3 home, the current across L3 home will be (-9A) – (+9A) = -18A which is higher than the 9A that home normally takes in so the equipment in that home will all burn. Power company staff reported to this author; that if an outgoing N wire broken by a lorry hitting it, will cause the home appliance to be burnt. Initially this author said it cannot be because the outgoing side of the three homes will be a star point like some induction motors. But as more of the power company staff reported the same thing this author realized it must be the current differential in the three homes that cause the appliances burning while in an induction motor the three coils use the same amps (balanced three-phase loads).

It is for this reason that power companies are always concerned about N being broken. A power company's Linesman's main job is to ensure the N is not broken anywhere in the system. One university engineer informed that a few whole labs of computers blew up due to a faulty N connection.

## 5.1 Synchronizing generators

When there is a need to combine two generators to provide more power than one can provide, they need to be synchronized. The schematic for synchronizing is shown below in Fig. 13.

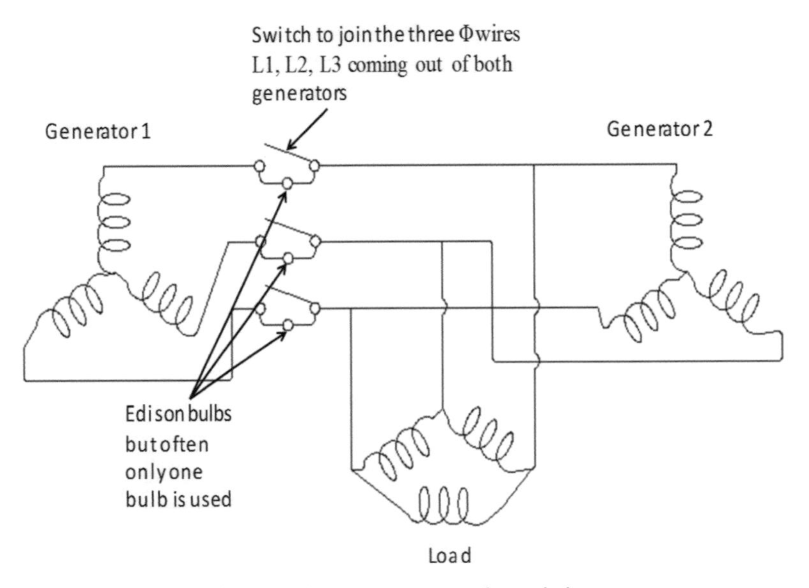


Fig. 13: Generator synchronizing

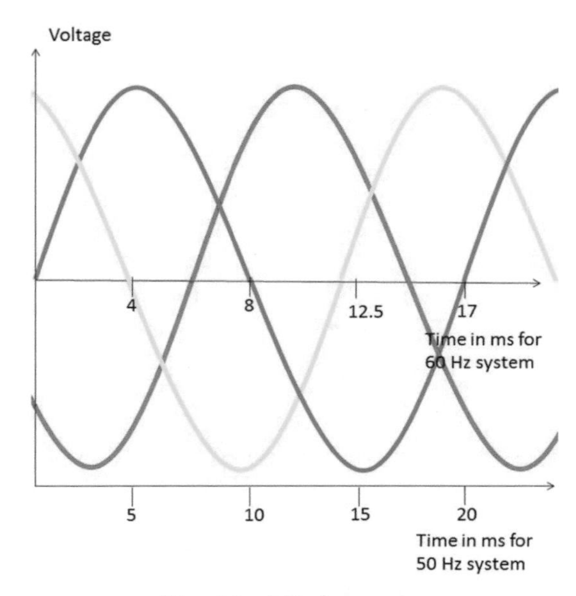


Fig. 14: AC sine waves

What happens here is that when say L1 (R) phase wire of generator 1 and L1 (R) phase wire of generator 2 are joined to the two terminals of an Edison bulb (incandescent bulb), it will light up if the two generators' waveform is not in phase, that is, just like connecting the incandescent bulb to L1 (R) and L2 (Y) phase of the first generator where the voltage will be 120√3=208V (240√3=415V). As can be seen from Fig. 14 at most moment of time (the X-axis) there is an amplitude difference between L1 (R) and L2 (Y) phase. This amplitude difference is energy to light up the bulb. But if the DC going into the solenoid rotor of generator 1 is played with (increased or decreased), the rotor will at some moment rotate at the same speed as generator 2 because the variation of current in the rotor will make the rotor more or less magnetic. When the rotor is more magnetic, it will turn slower across the stator coils (which have recently become electromagnets also because it is a coil with generated current flowing in it) and when it is less magnetic it will turn faster across the stator coils. The three wires of the stator carry the generated currents out to the loads. For example, if the DC is increased, the rotor's magnetic field will interact with the stator's magnetic field, to slow down the rotor thereby slowing down the AC waveform produced. So, the rotor can be slowed till the waveform produced reaches the speed of generator 2. At this moment the light bulb will go dark because the waveforms from the two generators will be in phase and L1 (R) wire of generator 1 and L1 (R) wire of generator 2 can be joined. Normally only one bulb is joined instead of three to confirm all three phases are synchronized. This is because in a rotating generator, if one phase is synchronized, the other two phases has to be synchronized.

Note this author has tried increasing and decreasing voltage to an Edison bulb and it doesn't blow up even if the voltage is varied from 0V to 415V. So, it is good to have at least one Edison bulb in a home because no other bulbs can stand such variation in voltage. It will probably be the only thing that can survive an EMP (Electromagnetic Pulse) attack.

# Chapter 6

## Understanding 1Φ and 3Φ power

In this chapter many times LV or Low voltage is referred as single-phase or three-phase. Single-phase voltage is 120V in USA and 240V in Britain. Three-phase voltage is 208V in USA and 415V in Britain. If the AC single phase voltage is X V, this is the RMS (root mean square) value of the voltage which when rectified will give about the same DC value as shown below:

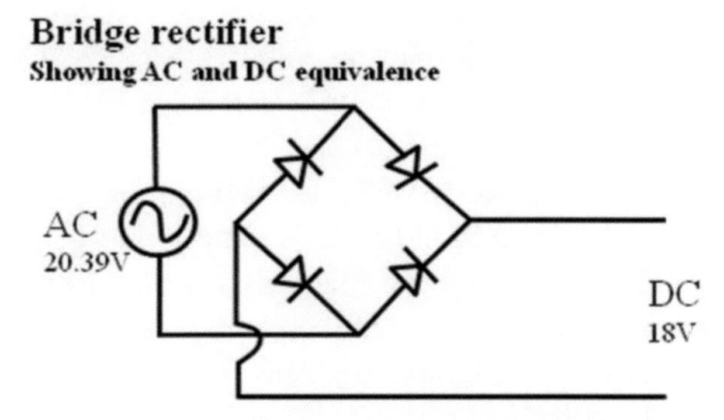


Fig 87.1: Bridge diode experiment, refer to Fig. 87.2 for full functioning

Experiment: Input AC from a transformer was measured to be 20.39V AC. After the bridge rectifier the voltage is 18V DC. 20.39-18=2.38V.

Theoretically a diode takes a 0.7V drop, so theoretically the drop should be 0.7 X 4 = 2.8V

Fig. 87.1 indicates the equivalence of $V_{RMS}$ AC and DC. The reason for creating this RMS value was to make AC equivalent to DC which was the original system created by Edison and company. Note that Edison's original system was mainly to heat up water at the bottom of buildings and to pump this hot water to all homes in a tall building. In each room of the building the pipe will be turned into a spiral which will replace the fireplace. This was done because a fire accident in a fireplace can cause a whole building to burn and many to die. This electrical system was a safer one for people. The actual measurement of equivalence can be made with an electric heater supplied with DC and another one supplied with AC. Say if single phase DC required 10 minutes to boil a bucket of water, it will take an AC peak voltage of (DC voltage / 0.707) to heat the same bucket of water in 10 minutes. The technicians of those days did not want to remember a new value of AC peak voltage so Tesla created the RMS = (peak AC voltage value X 0.707). The meters can only measure the peak AC voltage or $V_p$. In analog meters the scale of the AC meter is just changed to read 0.707 X Peak voltage to get the RMS voltage ($V_{RMS}$) value while in digital meter a $V_p$ X 0.707 is done to get $V_{RMS}$.

Why is 50 Hz used by most of the world? When the magnet is spun as a rotor and say there is a single coil in the stator. And this coil is connected to an Edison bulb. Before the rotor magnet passes the coil, the Edison will not light up. As the magnet passes the coil, the Edison bulb lights up. If the rotor is made to spin faster switching on and off, of the Edison bulb happens faster. Tesla found that at 55Hz the human eye cannot see the flickering. So, he set the USA frequency to 60 Hz. But when this technology was carried over to Germany, the Europeans did not want to follow a convention from the USA. Being supposedly more civilized, plus the fact that they preferred to be more metric. Since 50 is 100 divided by 2, they chose 50Hz instead. Europeans colonized most of the world so most of the world is using 50Hz. Japan had a closed-door policy and when they opened up; they decided to emulate the country best in the respective fields. So, Japan followed mechanical engineering from the Germans and electrical engineering from USA. Therefore, Japan is one other major country which uses 60 Hz (actually about half of Japan uses 50Hz). Other than Japan, countries close to USA from Brazil up to Canada use 60 Hz.

In a single-phase system, load has to be balanced in the housing estate or among machines in a factory such that each of the three phases have almost equal load (current drawn). That way the current in the neutral wire will be as low as possible. For example, an extremely unbalanced housing estate is L1=10A, L2=1A and L3=-2A.

Summing them will be 10+1-2=9A. So, the neutral will be carrying -9Ais carrying very much more current than the phase wires. If the figures for a housing estate is more real like +300-5+5 =300A, so the neutral wire at that moment of time is carrying -300A which could be over the current carrying capacity of the neutral wire, causing it to overheat and eventually burn. The overheating of the neutral wire is the prime reason for balancing the three loading in the three phases. This is one of the prime jobs of the Linesman in electrical power companies i.e. ensuring the three phase currents are balanced. They will go up the tower of overhead lines to reconnect power to homes in a housing estate to various phases such that each phase has about equal loading.

If a home itself has all three phase wires coming in, the loads within the homes should be balanced such that overall, each phase uses about the same current. Many fails to understand this and think three-phase home are totally different. The various equipment in a three-phase homes are single-phase. It is only in factories where all three phases are utilized to run machines. Most of such factory loads are three-phase induction motors. Rarely though, a rich person will install a 50,000 BTU air-conditioner, that is one of the very rare home equipment which need all three phases to run. So, in a typical three-phase home, the wiring scheme can be as follows: two bedrooms draw power from one phase, the kitchen draws power from the second phase and the front hall draws power from the third phase. Basically, an arrangement like this is done so as to balance the loading in all three phases. In single phase-homes, wiring scheme can be three homes using one phase the next three using the second phase and the next three using the third phase.

Two forms of connecting three phases are the Δ and the Y connection. In a Δ connected motor, neutral is not connected. In Y connected motor, neutral is usually connected (it can still run if N is not connected) because as previously mentioned in this book, there are three ways to achieve the other side of a load, or zero volts, namely, neutral, ground or Star point (joining all phase wires after the load).

AC going to a motor in Δ connection is analogous to three pipes sending water to a motor. The water is sloshing back and forth in each of the pipes. The motor uses this 'back and forth sloshing water' to do mechanical work. The phase to phase voltage averages at 415V but actually it varies from 0V to 294V (208/0.707) in USA and 0V to 587V (415/0.707) in Britain. So, in a three-phase induction motor two out of six coils will have a voltage which continuously varies from 0V to 294V (587V in Britain), thereby magnetizing and demagnetizing the coils according to that voltage variance. As two coil demagnetizes when 0V is across them, the next two coils get increasingly magnetized and so forth, therefore creating a rotating Magnetic Field (RMF) in the stator of the induction motor. Looking at Fig.15, the X-axis is time and the space between the blue and yellow wave is the magnitude of the magnetization on a set of two coils of an induction motor over time. Note that magnetism is caused by the current flow but as mentioned previously in this book, the current usually follows the voltage. But increasing voltage with little current cannot increase magnetization because there is no V in the formula that defines the magnitude of magnetization but the current is in the equation. You can think of it as I as the real thing that get things done, the V is just the force to move it. It should be noted that though the peak values of single phase of 120/0.707 = =170V (in USA and 240/0.707=339 in Britain) do not seem too much higher than the three-phase peak voltage of 208/0.707 = 294V (in USA and 415/0.707=587 in Britain), but the effect is very much different. In this author's technical school, sometimes student make wrong wiring and energize, for a single-phase short circuit, even next student who is just two feet away might not know it happened because it is so quiet. But when a student causes a three-phase short circuit, it can be heard in the whole 90 feet room separated into three, the lab, classroom and toilet. Normally the incoming cut-out fuse for the classroom will also blow up and must be replaced. So, do not think a three-phase short circuit is alright. This author has accidently touched 240V live wire probably about 300 times over his life-time but three-phase voltage wire is definitely something this author hopes never to touch.

In a single-phase motor, the coils magnetize following the sinusoidal wave. So, looking at Fig. 15, the blue curve. A coil in a single-phase induction motor get gradually more magnetized as the curve moves above zero, has zero magnetization as it crosses zero of the Y-axis and is appositively magnetized as it goes below zero and so forth.

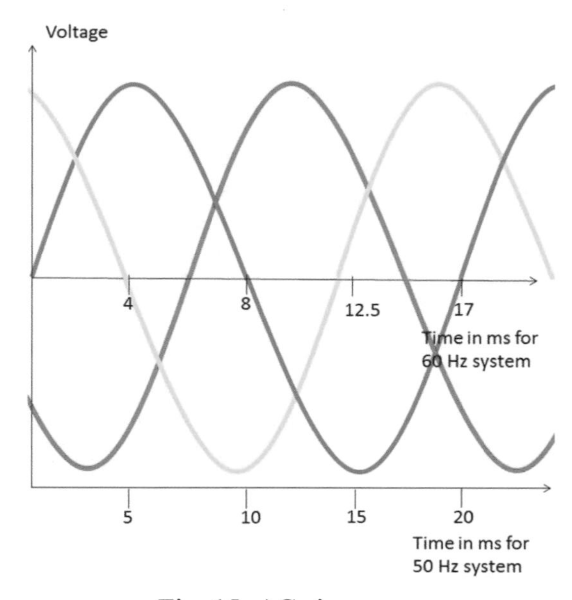


Fig. 15: AC sine waves

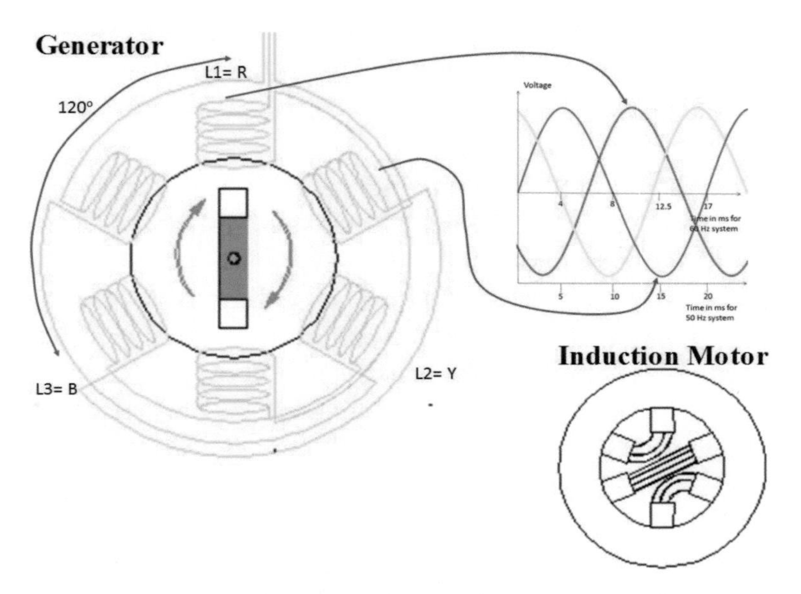


Fig. 16: Generator and induction motor winding

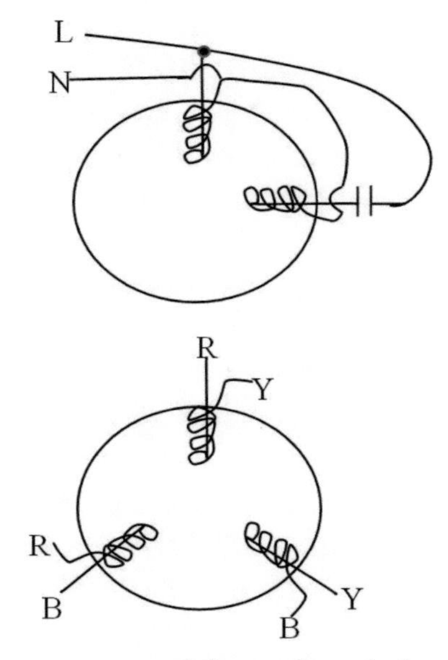


Fig. 17: Single phase motor on top and three phase induction motor at the bottom

As can be seen from Fig. 17 top and Fig. 15, a single-phase motor anticlockwise coil connected to L3 (B) phase will have zero magnetic strength at 0ms, a peak magnetic strength at 5ms and zero magnetic strength at 10ms. It will have opposite peak magnetic strength at 15ms and back to zero magnetic strength at 20ms.

Actually, as shown in Fig. 17, the single-phase motor is a modification of the three-phase motor. The modification is with the use of a capacitor. The top coil (assume a home fan) of a single-phase motor is energized with L3 (B) phase live wire. L3 (B) phase coil magnetizes as the N magnet rotor passes the L3 (B) clockwise coil in a 1000km away generator (assuming the coiling in the generator is exactly as the far away three-phase motor; it is not always the same. It is the same in steam-turbine powered generator but different in a gas turbine and hydroelectric power station). Later in the generator, as the magnetic N of the rotor passes the anticlockwise L3 (B) coil thereby magnetizing the fan's bottom coil. That is, initially the top coil is magnetized and later the bottom coil is magnetized. Note, while the rotor's magnetic N passes the top clockwise coil, the rotor's magnetic S passes the bottom anti-clockwise coil; when that happen the amplitude of the current generated increases by a factor of two. So, as the generator rotor's magnetic N passes the top coil, the top coil in the 1000km away induction motor becomes a N magnet and when it passes the bottom the bottom coil of the 1000km away motor becomes a N. Therefore, the N circulates around the stator of the single-phase induction motor. But the problem is, the top coil and the bottom coil of the single-phase motor are too far away to start the non-magnetic squirrel cage rotor of this single-phase motor to turn. To solve this problem, an additional pair of coils is installed horizontally in the stator of the single-phase induction motor. The same L wire goes to the top coil is jumped to a capacitor before going to the horizontal coil. The affect is that, as soon as a switch to this motor is switched on by a human, it will immediately energize the top set of coils but the horizontal coil will be energized only after charging and discharging the capacitor in-between. In-effect the horizontal coil will be magnetized slightly later. This is just right to start the rotor turning and eventually follow the rotating magnetic field (RMF) asynchronously. After a short while, a centrifugal switch cuts off power to the horizontal coil and the motor follows exactly the turning of the rotor of the far away generator. So, the purpose of this horizontal coil which is called the starting-coil is to give the starting torque to the single-phase motor and to give it direction; clockwise or anticlockwise. This author did try shorting the legs of the capacitor of a single-phase motor. The motor did not turn. But if a finger is used to tug the blade in the clockwise direction, the motor turns clockwise. If a finger tugs the blades in the anticlockwise direction, the motor will continue spinning in the anticlockwise direction. The problem with single-phase motors is the capacitor. Capacitors have a relatively short lifespan so single-phase motors cannot last as long as a three-phase motor with equivalent varnish around the coil, purity of copper in the coils and bearings. For a couple of years, this author was in charge of the repair center of the Western Digital factory, where all spoilt equipment was sent. In most cases, heat is the root cause of failures in electrical equipment and the heat normally destroy capacitors. Sometimes a $0.20 capacitor can bring up a $7000 VFD (Variable Frequency Drive). Another thing to note is that all electrical power connection to equipment or components of equipment worldwide are in parallel except the capacitor before the starting coil of a single-phase induction motor which is connected in series with the L wire. The above description of a single-phase induction motor seems like a small thing but single-phase induction motors are one of the most prevalent equipment worldwide, driving every home fan, air-conditioners, water pumps, and most factory pumps.

Polyphase phase motor was invented by Tesla prior to the usage of polyphase as an AC supply was standard. When Tesla first came to work for Edison in New York, he described his induction motor invention. Edison was aghast, stating that this is a totally wrong idea because the whole grid system which belonged to Edison's company in those days was DC and this motor only works with AC. That original invention was a three-phase induction motor. Therefore, three-phase motor is the original and meant to work well with the three-phase supply. Edison initially said, there is a missing part on this invention, a commutator (a pair of carbon brushes plus a copper ring with laterally cut out portions). This is because all motors of those days were DC with commutators. But Tesla said his motor runs on AC. Edison then said, "you want me to change my whole system just to run that small motor of yours". But today 68% of industrial electrical energy generated runs induction motors. This is one of the proofs that Tesla is right most of the time (probably always) but what he states may come true only after he has passed away. Some people do not understand this. For example, one Technical Manager of a big hotel to which this author was a contractor to wanted to change all the original three phase motors used in the hotel's kitchen blower and exhaust fans to single-phase motors. He probably felt the single-phase induction motor is simple because it drove his home fan and air-conditioner so he wanted to implement it in his hotel as well. This author advised him against that because three-phase motors gets its RMF rotating round the stator exactly in-step with the magnetic rotor in a 1000km away generator and it runs without a feeble capacitor. Also, the load in the three phases of the big hotel will be easier to be balanced with the major load being balanced-three-phase induction motors.

Studying this in detail, looking at the phasor diagram of Fig. 16, if an induction motor has three coils connected to (L2, L3) Y, B phases (L2, L1) Y, R phase and L3, L1 (B, R) phase. L2, L2 (Y, B) coil will have magnetic energy till 2 ms, L2, L1 (Y, R) will have magnetic energy till 5ms and L3, L1 (B,R) will have magnetic energy till 7 ms. Thus, as one coil loses magnetism as the phasor voltage waves meet, the other two coils will have magnetic energy to turn the rotor and so forth. As the magnet in the far away generator passes the L1 (R) phase coil, the coil in the motor in even 1000 km away will be a magnet almost immediately. The magnet then passes the B anticlockwise coil and immediately the B anticlockwise coil in the motor becomes a magnet. Then the generator magnet passes the Y clockwise coil making the clockwise coil in the motor a magnet. Thus, the rotor just follows the generator movement.

How fast can the current from the far away generator travel can be calculated approximately as follows, assuming current travels at the speed of light (in bare overhead lines it travels at 97% speed of light):

$$S = \frac{d}{t} \quad so \quad t = \frac{d}{s} \qquad (17)$$

Assume 1000km distance:

$$t = \frac{d}{s} = \frac{1000\,km\,X\,1000m}{3\,X\,10^{8}\,m/s} = 0.003\,seconds$$

Neutral is connected to ground at the incoming transformer which is why a test pen on neutral does not light up. But it is not safe to touch neutral because the three phases may not be perfectly balanced in places like a factory production floor. Say, L1 has a very high loading and L2 and L3 has very low loading, this will cause the voltage in L1 to drop. So, when L1, L2 and L3 voltages are added, it can be above 40V which is enough to enter a human body.

Ground wire (green or Earth wire) on the other hand is connected to copper rods all over the factory, hospital, airport etc., thus it is always at zero potential and safe to touch. The Fig. 18 shows the schematic of the intake transformer to a factory or home. This intake transformer is normally located in a compound near the electrical installation; like a housing estate in a fenced-up area called the Ring Main Unit (RMU) as shown in Fig. 19. In the RMU there is normally a transformer plus a switchgear.

Normally, an 11 kV line is already running underground. If a businessman needs to build a new shop house, the power company finds the nearest 11kV underground line and instructs where to locate and build the RMU. Therefore, the RMU is usually located on a 11kV line. The main switchgear has two individually controlled switches, that is one can switch ON while the other is switched OFF or both can be switched ON etc. Coming out of these two switches is a big ceramic fuse about 4" in diameter and about 14" in length.

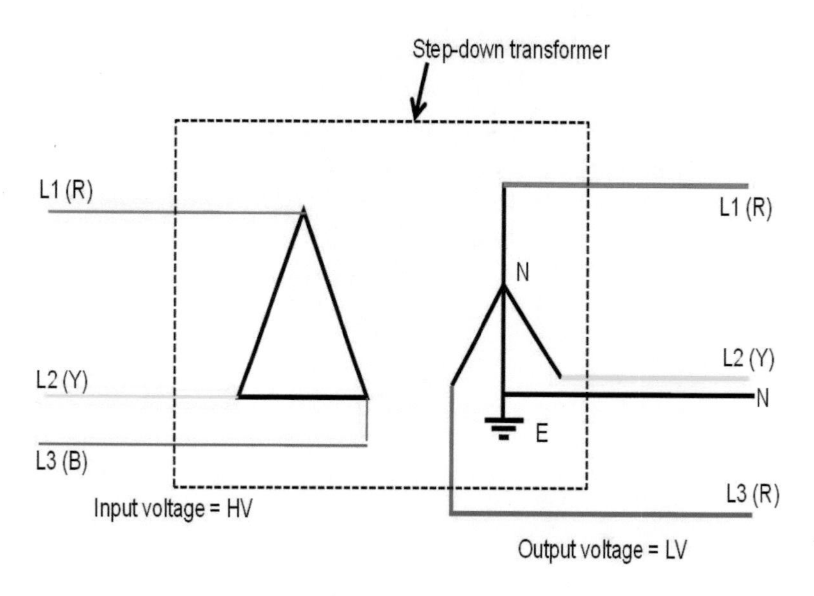


Fig. 18: Transformer substation schematic; typical layout to HV to LV substation

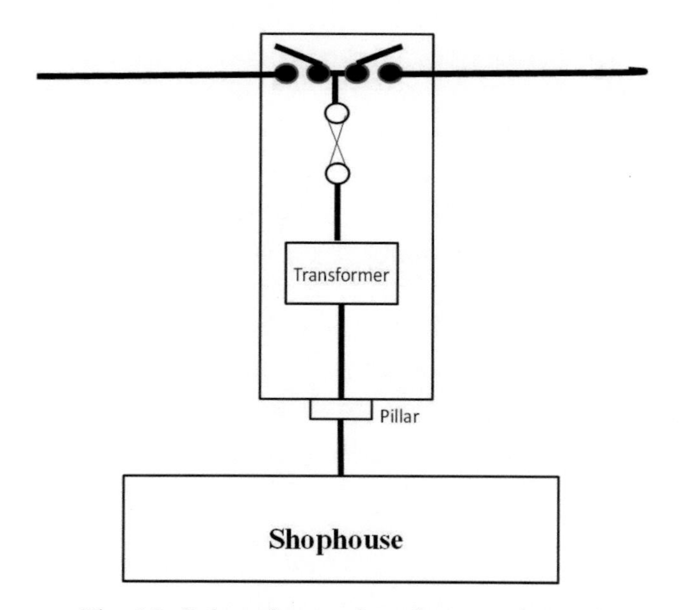


Fig. 19: Substation to shop-house schematic

In the transformer the out-going will have Star connected coils. The three coils L1, L2, L3 (R, Y, B) are looped as a star point and this star point is joined to the Ground (Earth) rod. Sometimes up to 40 Ground (Earth) rods are planted and all these Ground (Earth) rods are joined with a 1" X 1/8" copper tape or $35mm^2$ bare wire. This Ground (Earth) must be good because this is the return point of the all the three phase L1, L2, L3 (R, Y, B) live wires that goes to the load (home, factory etc.). For example, if this return path is not as good as possible, the current in the live wire will try to go to Ground (Earth) via human bodies as shown in Fig. 20. But if neutral is very good it will not at all go through a human body even if a human is touching a live wire. Remember electricity just have one 'mind', which is to just get to the Ground (Earth) the fastest way it can.

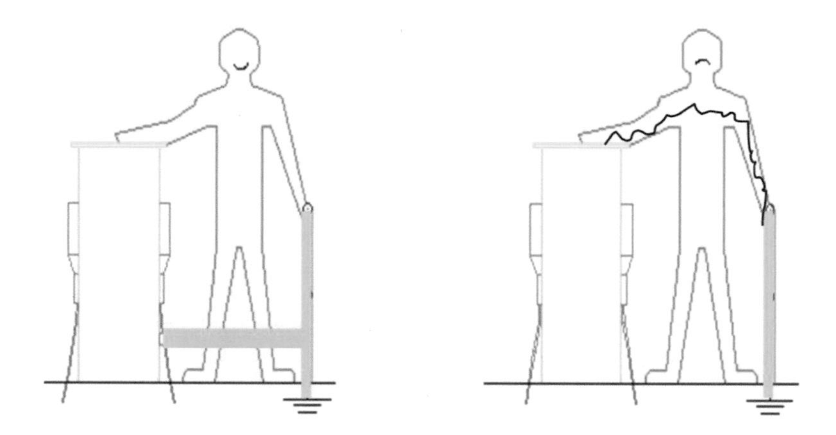

Fig. 20: Man touching live switchgear casing without earth bonding on left and man touching live switchgear casing with earth bonding

Current takes the easiest path to ground and will not at all be interested in going through a human body which has resistance. Normally substations transformers the Ground (Earth) resistance is below 1Ω sometimes specifies as 0.5Ω. Generator stations also need such low resistance. The resistance measurement is taken with a Ground (Earth) resistance meter as shown in Fig. 21.

Fig. 21: Ground (Earth) Resistance meter used to measure earth resistance of a telecommunication tower grounding (earthing) system

This meter has three wires, the green wire is clipped to the Ground (Earth) rod and the other two wires are factory built to a correct length to detect the Ground (Earth) resistance. Thus, these two wires need to be pulled as far as possible and at that end it is clipped to an eight-inch-long T metal (Fig. 22) which is pushed into the ground, easiest with a human foot. The two Ts are sold together with the meter.

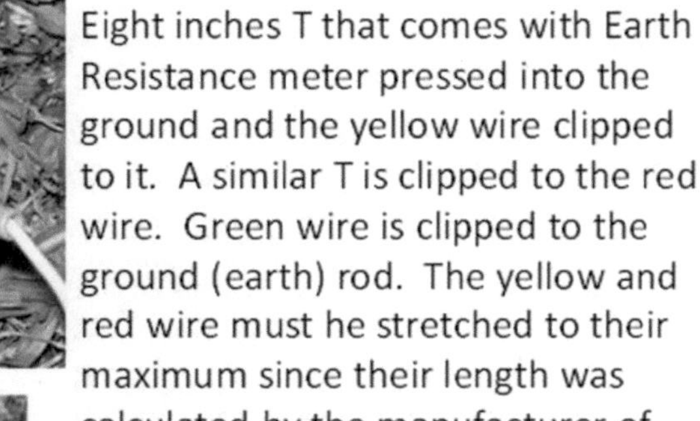

Eight inches T that comes with Earth Resistance meter pressed into the ground and the yellow wire clipped to it. A similar T is clipped to the red wire. Green wire is clipped to the ground (earth) rod. The yellow and red wire must he stretched to their maximum since their length was calculated by the manufacturer of the meter.

Fig. 22: Earth Resistance meter

The main orange button on the Ground (Earth) resistance meter is then pushed to get the Ground (Earth) resistance measurement.

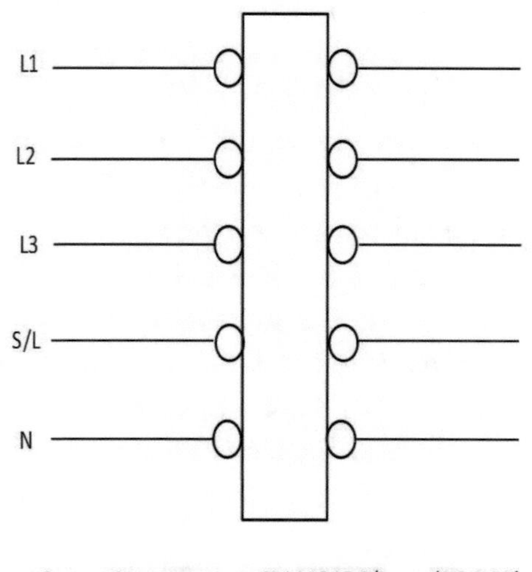


Connectingsequence: N, L1,L2,L3,S/L (1.2,3,4,5)
Disconnecting sequence: S/L, L3, L2, L1, N (5.4.3.2.1)

Fig. 23: Jumper connection in overhead lines

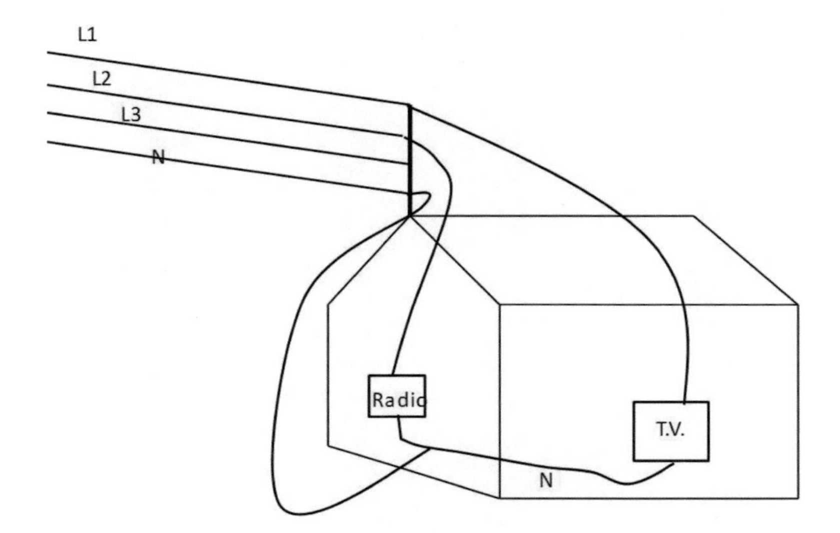


Fig. 24: Problem of not having N connection in three phase home

When connecting live jumpers across section poles as shown in Fig 23, the sequence to be followed is N then L1, L2, L3, S/W (R, Y, B, S/W). S/W stands for Switch Wire which is used to power the street lights. The reason for this is say, L1 (R) and L2 (Y) is connected as shown in Fig. 24, without N being connected. Now the TV in this three-phase home is connected to L1 (R) phase and the radio is connected to the L2 (Y) phase. Now L1 (R) and L2 (Y) are joined via the N wire giving $120\sqrt{3} = 208V$ ($240\sqrt{3} = 415V$ British) powering the TV and radio. The TV and radio are supposed to receive only 120V (240V British) so they will be blown. Basically, the White (Black British) N wire has become a live wire. So, with the procedure of connecting N first is designed to provide 0V in N wire first before L1 (R) and L2 (B) are connected thereby the TV and radio will not blow up. N is basically Ground (Earth) at the substation. In some Australian farms, some of which are up to 100km X 100km, it would be expensive to run L and N wire so only L is run to the house and after the load (say a TV) the return path is ground rod of the home. Basically, as mentioned before electricity only wants to go to the ground and if some equipment is placed in-between the source (generator) and Ground (Earth), that equipment will function.

# Chapter 7

# Line voltage and phase voltage

A three-phase system has three phase-to-neutral voltages that are most of the time equal in magnitude and displaced by 120°; because the generator coils in-between one phase's (say L1) clockwise coil and another phase's (say L2) clockwise coil are physically 120° apart. This will cause the top peak voltage of L1 to be 120° apart from the top peak voltage of L2 when observed with an oscilloscope. The voltages in each phase are often referred as **phase voltages** which is actually the phase-to-neutral voltage. The phase voltage is written as $V_p$ in Fig. 25 and Fig. 26 which are the two different methods of joining the phases up in a motor or a transformer.

Fig. 27 was drawn with excel. Going by L1, L2, L3 in sequence, the 1st coil should be L1, L2 then L2,L3 then L3,L1. But as explained in Fig. 37, this will not work, the phases need to be flipped on the third coil so the wires going into the three coils from top are, L1,L2 then L2,L3 then L1,L3. This is because it won't work for the outgoing of the second coil to be L3 and the incoming of the third coil to also be L3. Also, the outgoing of the third coil cannot be L1 and the incoming of the first coil to also be L1. But installers of induction motors do not need to bother about this as this flipping is already done by the manufacturers. So, there are six terminals of an induction motor, three up and three at the bottom. Installers just have to connect L1,L2,L3 on top and L1,L2,L3 at the bottom to get it running.

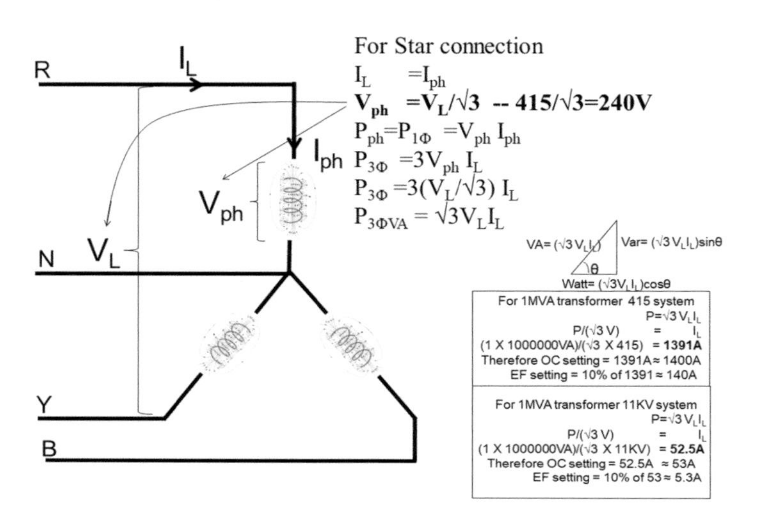


Fig. 25: Star connection

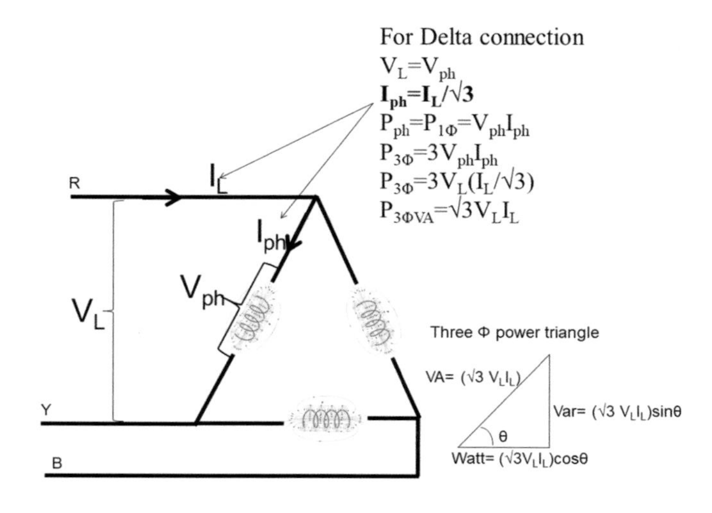


Fig. 26: Delta connection

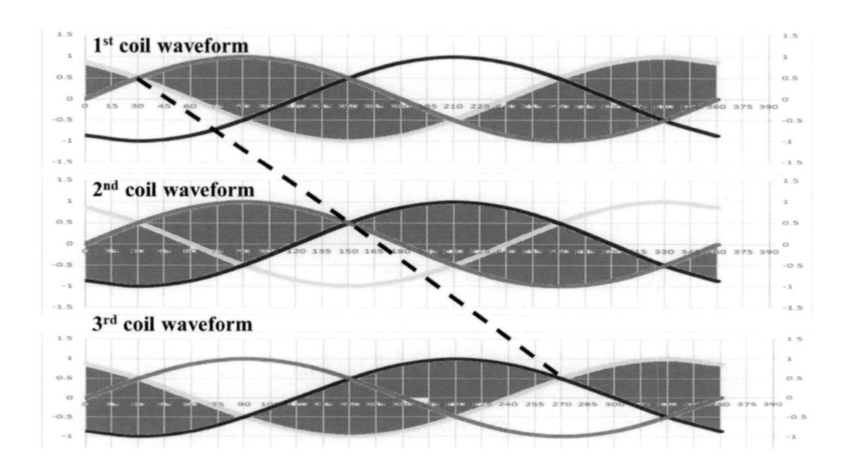


Fig. 27: Three-phase waveform going into a two-pole three-phase induction motor

For a 60Hz American system, the motor turns at 60 cycles---1 sec, =>1cycle---1/60=0.0166=16.6ms for a full cycle. It takes 16.6ms for a full cycle in 60 Hz system, so, 60°---16.6ms => 240°---16.6/360 X 60 =2.766ms between one coil of an induction motor deenergizing to the second coil deenergizing.

For a 50Hz British system, the motor turns at 50 cycles--- 1 sec, => 1 cycle---1/50=0.02=20ms. It takes 20ms for a full cycle in 50Hz system, so, 360°---20ms => 60°---20/360 X 60 =3.33ms between one coil of an induction motor deenergizing to the second coil deenergizing.

# Chapter 8

## Protection, Conductors and Insulators

The function of electrical protection is mainly to safeguard the conductors because if conductors burn, homes or buildings will burn. Copper is the most common electrical conductor, with aluminum catching up fast. It has a melting point of 1083 centigrade, an atomic radius of 1.57 Angstrom ($10^{-10}$ m). The outer shell of any atom is the valence shell, for Cu the valence shell has one electron as shown in Fig. 28. But one more electron from the next shell always like to be free also. So copper atom always like to release two electrons. These two electrons are free which makes copper a conductor. So, a piece of copper has twice as many free electrons as atoms. Insulators atoms like plastic likes to take into their atom an electron or more so there are no free electrons to conduct electricity. So, if a positive wire and a negative wire is placed over a length of copper creating a voltage (a force) between these two points, the free electrons within it will flow. But if the same voltage is placed across two points of a piece of plastic like PVC, there are no free electrons to move from negative to positive. But is should be noted that there is such thing as a 'breakdown voltage' for any material, over which there will be flow of electrons. Meaning if sufficient voltage is provided, anything can conduct. This is equivalent to the fireman stating there is a "flash point" at which any anything can burn, even cement. This information is critical in understanding when doing live work, it is not just an insulator plastic which protects you but how thick or what type of plastic or what is the plastic's breakdown voltage. The most common term used today is dielectric strength. The word dielectric was originally used for the material in-between capacitor plates. The higher the insulating property of this material, the higher the capacitive value of the capacitor. So today dielectric strength is used to replace the words electrical insulation capacity.

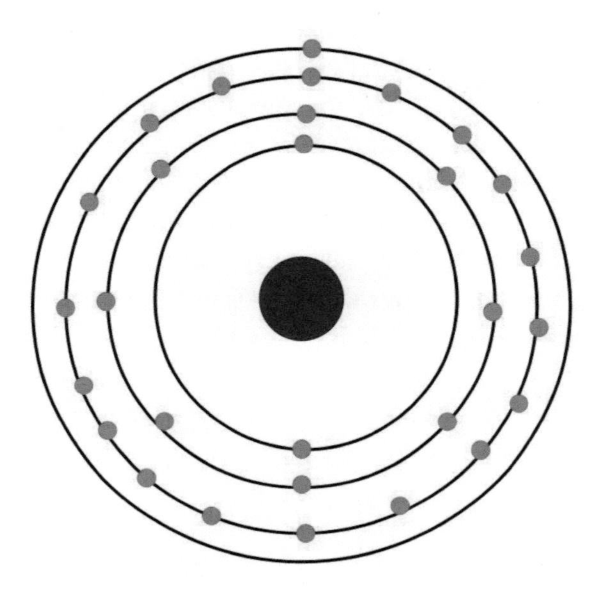

Fig. 28: Atomic structure of copper

With the help of STM (Scanning Tunneling Microscope), humans can actually see the atom today. The inventor of the STM received a Nobel Prize because it brought all of humanity one step ahead in being able to see the atoms which were theorized long ago.

An ion is an atom or molecule in which the total number of electrons is not equal to the total number of protons, giving it a net positive or negative electrical charge. An **anion**, from the Greek word ano, meaning 'up', is an ion with more electrons than protons (**negative charge**). A **cation**, from the Greek word kata, meaning 'down', is an ion with more protons than electrons (**positive charge**). Another way to remember is that the **positive cation moves to the cathode** to receive electrons

It should be noted that all ions, even ions of plastics are conductors of electricity. Gas are insulators but ionic gases are conductors. Note plasma, ions, and free-radicals (as called by doctors) are the same thing. So, ions should be kept away from electrical systems as far as possible. In a MSB (Main Switch Board) room, the specification states there should be a window to bring in atoms and let the ions. The ions are created especially at the capacitor bank section of the MSB and these should diffuse out of the room. The specification also states that if the room is underground there should be a fan at this window (forced ventilation). MSB rooms create lots of ions especially at the capacitor bank. Here, high-powered capacitors are continuously being switched on and off according to the amount of inductive load present; resulting in arcs in the high-power contactors. This continuous switching of MCCBs is done automatically by the Power Factor Regulator (PFR) to correct the power factor (PF) of the installation. In the MSB, before each capacitor is connected a MCCB (Molded Case Circuit Breakers) and then a high-power contactor which receives instruction from the PFR. MCCB handles currents from >100A to 1000A. A MCB (Miniature Circuit Breakers) of homes handle current from 1-100A. The arcs occurring in the contactors as it switches on and off convert atoms of the air in the MSB room into ions. Basically, the industrial method of creating ions is to blow atoms over an arc. Note only a person having a Chargeman license can legally enter the MSB room, so it is possible nobody opened the MSB room door for a month or two. By regulation and for safety this door must be shut. So, if there is no ventilation, eventually the MSB room will be filled with ions. Then one day, when there is a fault in the building and the ACB (Air Circuit Breaker) need to operate, conductive ions instead of insulator atoms will be blown in-between the ACB contacts resulting in an explosion.

Some MSB rooms are designed with air-conditioning which just recycles the existing air. Electrical and electronics in the MSB last longer with air conditioning but even a small 1' X 1' window is essential to allow the ions to escape to the atmosphere and new atoms to come in. The window of the MSB room should be covered with a netting preferably metallic netting so that animals especially lizards cannot go in. This author has seen holes made in fiber glass nettings made by lizards, so, metallic or stainless-steel netting is necessary. This author also did a wiring job in a goat farm in a rural area where every single live point has a dead lizard. Lizards seem to like the relative warmth of live points. The Distribution Board (DB) needs to be totally sealed up especially in a rural setting. Lizards seem to be the most hazardous animal for electrical problems.

Note, these ACB are quite expensive (about the price of a Tesla model 3) because they are finely designed to quench arcs. But if conditions are slightly out of specification, it will explode. The method of arc quenching is by blowing insulator atoms over the arc, this will lengthen the arc and make it thinner and therefore safer. This blowing will also cool the arc a little. Note, creating ions is by blowing atoms over an arc and turning ions back into atoms is by cooling them with methods such as blowing insulator atoms over them. There was an accident where a 11kV switchgear was switched with a little low insulator oil within it. What happened inside this breaker was that the arc was blown but with less oil, therefore the length of the arc was shorter than specified. This caused an explosion which resulted in the switching personnel having second degree burns. That particular switching personnel got a 'Warning Letter' also after that because he was not wearing the required PPE. This is a problem because PPEs are designed for cooler climates and humans tend not to be able to withstand the heat within it in warm climates. Someone needs to invent an air-conditioning system within PPEs. Anyway, it is very critical that the breaker works exactly as specified, oil level checked and all the checklist of switching followed exactly

## 8.1 Miniature Circuit Breaker (MCB)

This idea of blowing an arc is used from the smallest MCB (Miniature Circuit Breaker) in the home to the largest breaker of big power companies. Fig. 29 depicts a typical home MCB. MCB has two protection system one is to protect against overload and the other to protect against short circuit. Say the above MCB is rated as 6A. Say 7A current flows in the L because the motor being run out of this MCB has a bearing jam problem and draws more than the specified current of 4A. This 7A will flow from the L wire into the left terminal (Fig. 29) of the MCB, it then flows through the solenoid wire. The solenoid wire is of big enough size such 7A does not turn the coil into a strong enough magnet. The solenoid is made with very thin wire so as to have some resistance (R).

**Functioning of a MCB**

Number written on the MCB like 4000A uses this system. It takes care of short circuits (S/C) and lightning strikes. A S/C or lightning strike is very high in current but as long as it is less than 4000A, the solenoid magnetizes and the plunger moves the white plastic, causing the MCB to trip immediately by spring action.

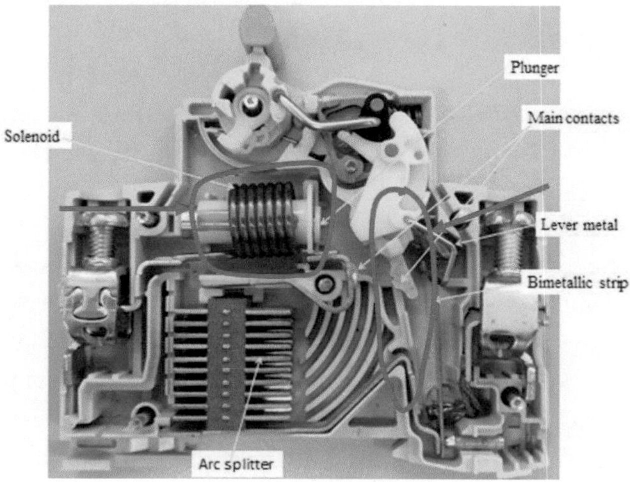


Number on the MCB like 6A is using this system. It takes care of overload (O/L). Say 7A flows through this MCB, the bimetal bends moving the lever metal which triggers the MCB to trip by spring action. Tripping happens slowly because the bimetal takes time to bend.

Fig. 29: Miniature Circuit Breaker (MCB)

But current will not be short circuited through it because as soon as it is energized, it provides reactance to reduce current flow. A 22-gauge copper wire has a resistance of about 0.016 Ohms per foot. So, 100 turns solenoid will have a resistance of about 1.6 ohms. Applying 12V DC to this coil will allow a current flow of

7.5A. If the wire is 28-gauge, 100 turn will give a resistance will be about 6.5 Ohms, and the current will be 1.8A. In MCBs the solenoid is made of much thicker wire so only a very high current from a short circuit can turn it into a solenoid. The current then goes through the silver-plated contact switch and down through the bimetallic strip and then goes out the left terminal. The bimetallic strip is a fused together copper on the right and iron on the left. Since copper melts faster than iron, it will bend and move the shiny lever-metal a little. This little movement causes the spring to suddenly trip the breaker, opening the silver-plated contacts. All circuit breakers, like this small one for a home, to the large ones used in electric power companies use a small signal such as this small bimetallic strip movement to activate a fast and strong opening of the switch with a spring. This is how overload protection occurs in the MCB.

If a short-circuit happens, a large surge in current occurs. This is because L and N wire are touching. And the only resistance when this happens is the very low one of the copper wire. V=IR, so, I=V/R. Say the V=240V and R of copper is 0.02Ω, so, I= 240/0.02=12000A. This is the reason why the current in a short-circuit is very much higher than in an overload. It is also why all breakers like this one for a home to the largest one used by a power company has two current ratings. For this MCB, the O/L current rating is 6A and the S/C rating is 4500A. This 4500A means when a 4499A short circuit occurs it can still operate. If a short-circuit of say 7000A occurs, this MCB is not rated to handle it.

When a short-circuit occurs, current will enter the left terminal, go into the coil and immediately turn it into a solenoid, causing the plunger to be pushed out hitting the white plastic and moving it a little which will trigger the strong spring to break the breaker immediately.

As the silver-plated switch opens, the mechanical design of the MCB blows air in-between the silver-plated contacts causing the arc to be sprayed to the arc-splitter below, thereby splitting and quenching the arc. Note the three mechanisms of arc-quenching which are by bowing the arc which will make it thinner and safer, then this thin arc is split by the arc-splitter and thirdly the blowing of atoms across an arc cools it; turning ions back into atoms. The other end of this thin arc need not go to the other silver-plated contact; it can finish up at the copper below the arc-splitter, thereby not damaging the silver of the silver-plated contact. Note the copper below the arc splitter goes all the way to the left because the arc which may happen in a fraction of a second could reach somewhere there due to the 120V (240V British) potential at this point. The arc emanating from the first silver-plated contact will not damage this copper compared to an arc shooting at the right silver-plated contact. It is for this reason it is better that the arc shoots at the copper below. The current then goes through the bimetallic strip. It takes about 30 seconds to a minute for the bimetallic strip to bend. So, in this short span of a short-circuit current, it will not bend. The bimetallic strip just acts as a conductor, as half of it is copper. Current then goes to the right contacts and on to the load. But this short-circuit current only occurs for a very short time because the MCB will break the circuit. Most loads can handle a very short span short-circuit or has extra protection such as HRC (High Rupturing Capacity) fuses to handle such surges. As the names implies HRC can are designed to 'Rupture' even if handle 'High' current flows through it. A normal fuses will allow quite a bit of high and fast current to go to the equipment before breaking.

When purchasing a MCB there are three categories, Class A, B and C representing 5X, 10X and 20X short circuit triggering respectively. Class A enables tripping at 5X the overload rating on the MCB. For example, a 6A rated MCB will trip with a short circuit of 5 X 6 = 30A flows through. Class B will trip with a short circuit of 10X (10 X 6 =60A) and a Class C will trip at 20X (20 X 6=120A) rated current. Class A, B and C are only related to short-circuit but the overload mechanism for all three classes are the same. Of course, the bimetal for a 40A MCB is thicker than a 6A MCB. The coils of the solenoid only have a few turns for a 40A MCB and many more turns in a 6A MCB. 40A – 60A is a huge current to pass this simple MCB; note 1A is enough to kill a human.

The reason for Class A, B and C are for example Class A is meant for light bulbs. An Edison bulb filament is cool initially so the atoms (or ions) are not vibrating much so the resistance is low. Note if the ions are not vibrating at all, there will be zero resistance as in a superconductor. As explained earlier in this book, resistance is proportional to the rate at which ions vibrate. So current can easily flow through the cable resulting is a very short period of time of almost short circuit. We do not want the MCB to trip because of this. Note the tiny span of time of very high ampere will not bend the bimetal of the MCB; it needs about 30 seconds to a minute of current flow to significantly bend the bimetallic strip enough to move the lever metal. Class B are for motors where inrush current upon starting can be many times steady state (running) current. Class C is meant for big motors where inrush current is proportionately larger.

From a good quality MCB all the way to giant switchgears, the contact points are silver plated (silver being the best conductor). This author uses Schneider because the three bigger electrical part supplier, GE, Siemens and ABB do not have an office in his hometown. Schneider has an office and at least can inform customers which electrical shop is selling original Schneider products. The office itself does not sell the products. There are imitation Schneider brand electrical switches in this author's home town being sold by other electrical shops. But as an electrical contractor, this author never sacrifices in brand. A world top company has a reputation to keep. And a contractor cannot blame the parts when a problem occurs. Therefore, it is the contractor's job to ensure the parts are of the very top quality. There are some very terrible brands out there in the market, often a MCB is just for show and do not do any work. So, the job of a good electrical contractor is to continuously keep up with who are the best international suppliers and also who are the best local suppliers. Big international suppliers do fail over time. Case in point is Enron or Arthur Anderson both of which do not exist anymore despite being world giants prior to scandals. For local suppliers, it is the experience of this author that if the owner of the shop is still in the shop, it should be more reliable. Owners, who dare to face disgruntled customers by being in their shop and not delegate everything to sales personnel, will more likely sell genuine electrical products.

The latest breakers blow sulfur hexafluoride ($SF_6$) in-between the contacts. In most common, Oil Circuit Breakers (OCB), the oil can eventually get ionized with many switching. Note sending an arc through an element or molecule is the method of converting atoms to ions. Cooling the ions is the method of changing it back to atoms. Hence the high-pressure oil in-between the contacts will cool the region and convert some of the ions formed back into atoms. The ionizer air-conditioner used in some homes has a small arc through which air is blown to create ionized air. Therefore, eventually the oil in OCB will be ionized and instead of blowing insulator oil in-between the contacts, a conductor (ionized oil) is blown resulting in an explosion will occur. It is very critical that the blowing of the arc happens with sufficient pressure and there is an arc splitter to quench this arc; every portion of this process must occur or else there will be an explosion. As already described earlier in this chapter, a technician switched a 11kV switchgear despite knowing the oil level was a little low, gave him second degree burns all over his body. This precaution in oil level checking is analogous to a car engine; the car cannot run without the engine oil. This author had an experience of driving over what seemed like a book on the road, which turned out to be a piece of sheet iron folded as a book. This sheet of iron punctured a hole in the engine oil filter but not having realized this continued driving a few miles, then the engine stopped dead. When the mechanic opened up the engine, there was massive damage all over the engine. All parts except the engine block had to be replaced; the engine block had to be re-bored. So, car engines simply cannot work without engine oil, just as ionized oil cannot work in an OCB. When this author worked for a power company all the blown-up OCBs were displayed as a warning to technicians to take precautions when operating them. The metal body of the breakers were 5-7 mm thick (like the iron of military tanks) but when it blows up it is twisted like thin sheet of paper; such is the power of electrical explosions. No one should take high power electricity lightly.

The correct procedure for technicians is to drain out a little oil at intervals specified by the manufacturer. Typically, it is once in five years. Most staff of power companies use time-based PM (preventive Maintenance) but it should be after a certain number of switching. Probably the hassle with record keeping is a deterrent to this procedure. The more times arching occurred through the switchgear oil, the more often the oil should be tested for ionization level; so, a better frequency should be the number of times switching occurred.

The equipment to test ionization level of the oil is called a Dielectric Tester. Higher ionization levels of the switchgear oil the less the dielectric strength or insulation capability. The word dielectric which originated as the material in-between plates of capacitors is now generally used to replace the word insulation strength in electrical power engineering books and journals. So dielectric=insulators and arc=sparks are lingua franca of electrical engineering.

The act of blowing air, $SF_6$ or oil in-between electrical contacts does three things:

1) The arc gets thinner.
2) The arc is split by many window-like insulators called arc-splitters at the outlet.
3) The material (oil or gas) blown in-between the contacts gets deionized by the cooling action of the blowing.

It should be noted that blowing atoms or compounds in-between arcs is the industrial way to create ions in the first place. Ionizer air-conditioners or general home ionizers have air being blown across arcs to produce ions. This author does not recommend either of these two products because, just as ions are dangerous in enabling random current flow in wrong directions, ions are very bad for the human body. Doctors call this free-radicals. If there are too much free-radicals in the body it gets damaged. This author was a Radiation Protection Officer (RPO) for 22 years and all radiation experts will state that it is not the $\alpha$, $\beta$, $\gamma$ or X rays that injure humans it is the second-effect of these radiation of kicking out electrons or adding electrons into atoms (which is basically termed ionization) that kill humans. Sales people of ionizing products state that it will keep rooms or refrigerators free of germs but the human body has more bacteria than human body cells, most of which are good bacteria. And what kills bacteria can kill human cells too.

In the compound $SF_6$, the sulfur and the six fluorine atoms have a very strong bond. Thus, $SF_6$ is unreactive and even very high arcs cannot split the F from the S, to ionize it. Sulfur and fluorine are easily available and cheap elements so countries that make $SF_6$ need not be advanced countries, anyone can take easily available sulfur (the earth's atmosphere had a high percentage of sulphur during the dinosaurs' time) and fluorine and heat it up in a container to make $SF_6$. It is harmless to the human body because it is so unreactive (it is a content compound) but as a global warming gas, it is 22,800 times more dangerous than $CO_2$; having a lifetime in the atmosphere of 800–3200 years. Therefore, the United Nations is watching out for the usage of this gas. The power company this author worked for now uses lots of $SF_6$ breakers from ABB and Siemens and the technicians report that the $SF_6$ do leak out. This made this author conclude that human technology is not capable of placing a gas in a container for long periods of time. This conclusion can be made because

Siemens is the world's number two electric equipment supply company and ABB is the world's number three electric equipment supplier company (GE is number one). If these over-100-years-old company cannot keep gas in a container, who can? Maybe gas can be kept in a fully enclosed container but in $SF_6$ breakers, there will always be a shaft to trigger the action of switching on or off and the O rings around these shafts must be the route for the $SF_6$ to leak out.

The very latest RMUs developed by Subhashish Bhattacharya and team of NC State University, USA as a collaborative project by the university, CREE and ABB. This RMU is very small due to its utilization of the latest development in IGBT (Insulated Gate Bipolar Transistors). The IGBT can switch at 700 ns for a 6.5kV and 1070 ns for a 10kV one.

Note the following data:

IGBT switching timing is 1070ns = **$1.07 \times 10^{-6}$s**.

Oil Circuit Breaker (OCB) switching timing = **$1 \times 10^{-1}$s.**

Thus, the IGBT **switches 100,000 times faster** than the OCB. It is because of this that IGBT used as switchgear can by-pass the requirement to blow a liquid or gas in-between the contacts. This causes the redundancy of lots of arc quenching parts thereby making the switchgear very small.

A 1 MVA transformer in this RMU is only 25" X 10" X 15" (250mm X 240mm X 390mm). This small size is enabled because this transformer is operated like a SMPS (Switch Mode Power Supply) where the higher the incoming frequency the smaller the transformer. The main advantage of this system is switchgears will not explode and kill people anymore. This author has a close Malaysian friend while studying at South Dakota State University, USA. He graduated and became an engineer in the main electric company in Malaysia. A year later he and three others were killed because the 11kV switchgear he was operating blew up. This IGBT RMU can prevent such fatal accidents. Other obvious advantage for places on earth where land is very expensive is that land need not be purchased to locate the RMU. The RMU is so small that it can be located in a small section of a home within the housing estate, which can be rented by the power company from one of the home-owner.

The power company statistics indicate the biggest culprit in tripping of overhead lines are squirrels because they are straight and long between the nose and the tail. Sometime snakes do climb up power lines and monkeys get electrocuted as they hold two phases overhead wires.

## 8.2 Residual Current Device (RCD)

The other protection device used in homes is the RCD (Residual Current Device). Previously this was called GFCI (Ground Fault Circuit Interrupter) or ELCB (Earth Leakage Circuit Breaker). While the GFCI (ELCB) detects fault current in the Ground (Earth) wire, the RCD detects a difference in the current value in L and N wires as shown in Fig. 30. The L wire will have a magnetic field around it following the Right-Hand Thumb Rule. The N has returning current and therefore has a magnetic field turning in the opposite direction compared to the L wire. These two magnetic fields oppose each other resulting in no current generation in the detecting CT (Current Transformer) coil. Note magnetic field creation around a wire is directly proportional to the current or electron magnitude. Voltage does not play a part in magnetic field creation. If a person touches a L wire, he will leak some current to the Ground (Earth) via his feet resulting in less current going back via the N wire. Therefore, the magnetic field of L wire will be higher than that of the N wire. The remaining magnetic field will energize the CT (Current Transformer) coil around both L and N wires. The word **Residual** of RCD refers to this **remaining** magnetic field created by the remaining AC after the N has neutralized the ones in L which is AC so it will generate a current in the CT. Current from this CT and will trigger an amplifier circuit to energize the solenoid which will break the circuit. Single phase RCD are set to trip at 100mA difference between L and N.

This situation can be compared to car workshop where air from a pump goes into the mechanic wrench as he opens a tire bolt. The air goes into a turbine that operates the wrench. The pressure of the air going into the wrench and out of the wrench must be the same but the work of turning the wrench is done in-between. So, in this electrical case, the current in L and N must be the same but lots of electrical devices are operated in the home in-between.

Another example is home powered by water pressure. A water pipe sends water to a home and another pipe sends the water out of the home. In between many turbines are tuning in the home energizing various equipment. But if there is a faucet in the home, there is a leak to ground (earth) and now the outgoing pipe will have a different pressure compared to the incoming pipe; this is equivalent to a human touching a live wire, draining some current into ground via his leg causing the RCD to trip.

Fig. 31 is a three phase RCD. In this case, the current additions of the L1, L2, L3 (R, Y, B) phases is -5A +3A -4A = -6A. So, the outgoing N wire will carry +6A. A CT placed surrounding all four wires detects nothing, because +6-6=0A. But if a human touch any of the L wires, he will leak say 1A current to ground (earth) through his feet and now the N wire will carry -5A. So, the combined phase wires into home will carry +6A and the neutral wire carries -5A. The CT will then detect +6-5 = 1A and this will be amplified a little to trip the RCD. Three phase RCDs are set to trip at 300mA difference between the three-phase combined (which goes into the home) and N (which comes out).

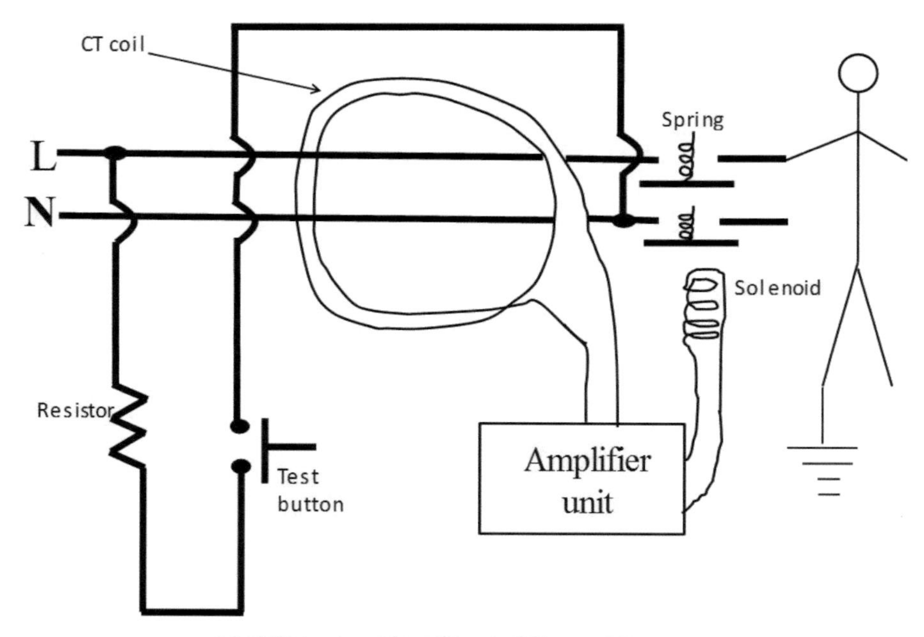


1Φ RCD is set to trip at 100mA difference between L and N

Fig. 30: Single Φ RCD schematic

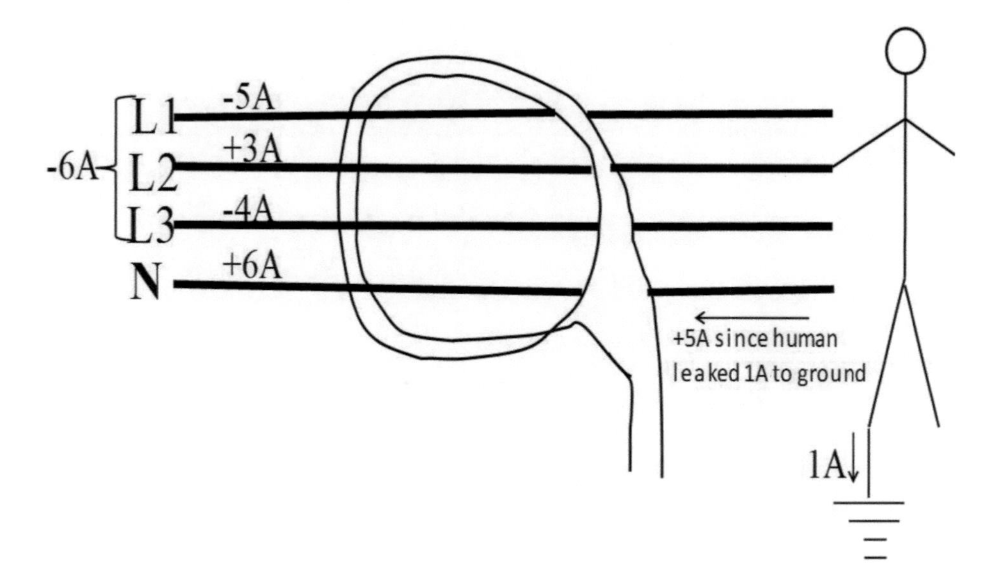


3Φ RCD is set to trip at 300mA difference between L1, L2, L3 combined and N

Fig. 31: Three Φ RCD schematic

Table 1: Resistivity of various conductors

**Resistivity of some metals**

| Material | Resistivity (ρ=Ω·m) at 20 °C | Melting point | Oxide melting point |
|---|---|---|---|
| Silver | $1.59\times10^{-8}$ | 962°C | |
| Copper | $1.72\times10^{-8}$ | 1083°C | 1200°C |
| Gold | $2.44\times10^{-8}$ | | |
| Aluminium | $2.82\times10^{-8}$ | 660°C | 2000°C |
| Calcium | $3.36x10^{-8}$ | | |
| Tungsten | $5.60\times10^{-8}$ | | |
| Zinc | $5.90\times10^{-8}$ | | |
| Nickel | $6.99\times10^{-8}$ | | |
| Iron | $1.00\times10^{-7}$ | | |
| Tin | $1.09\times10^{-7}$ | | |
| Platinum | $1.06\times10^{-7}$ | | |

**Soldering lead is 60% tin and 40% lead, Si boiling point is 2357°C**

Table 2: The Periodic Table

| 1<br>H<br>1.008 | | | | | | | | | | | | | | | | | 2<br>He<br>4.003 |
|---|---|---|---|---|---|---|---|---|---|---|---|---|---|---|---|---|---|
| 3<br>Li<br>6.941 | 4<br>Be<br>9.012 | | | | | | | | | | | 5<br>B<br>10.81 | 6<br>C<br>12.01 | 7<br>N<br>14.01 | 8<br>O<br>16 | 9<br>F<br>19 | 10<br>Ne<br>20.18 |
| 11<br>NA<br>22.99 | 12<br>Mg<br>24.31 | | | | | | | | | | | 13<br>AI<br>26.98 | 14<br>Si<br>28.09 | 15<br>P<br>30.97 | 16<br>S<br>32.07 | 17<br>CI<br>25.45 | 18<br>Ar<br>39.95 |
| 19<br>K<br>39.1 | 20<br>Ca<br>40.08 | 21<br>Sc<br>44.96 | 22<br>Ti<br>47.88 | 23<br>V<br>50.94 | 24<br>Cr<br>52 | 25<br>Mn<br>54.94 | 26<br>Fe<br>55.85 | 27<br>Co<br>58.93 | 28<br>Ni<br>58.69 | 29<br>Cu<br>63.55 | 30<br>Zn<br>65.39 | 31<br>Ga<br>69.72 | 32<br>Ge<br>72.61 | 33<br>As<br>74.92 | 34<br>Se<br>78.96 | 35<br>Br<br>79.9 | 36<br>Kr<br>83.8 |
| 37<br>Rb<br>85.47 | 38<br>Sr<br>87.62 | 39<br>Y<br>88.91 | 40<br>Zr<br>91.22 | 41<br>Nb<br>92.91 | 42<br>Mo<br>95.94 | 43<br>Tc<br>98.91 | 44<br>Ru<br>101.9 | 45<br>Rh<br>106.4 | 46<br>Pd<br>107.9 | 47<br>Ag<br>112.4 | 48<br>Cd<br>114.8 | 49<br>In<br>118.7 | 50<br>Sn<br>121.8 | 51<br>Sb<br>121.8 | 52<br>Te<br>127.6 | 53<br>I<br>126.9 | 54<br>Xe<br>131.3 |
| 55<br>Cs<br>132.9 | 56<br>Ba<br>137.3 | 71<br>Lu<br>175 | 72<br>Hf<br>178.5 | 73<br>Ta<br>180.9 | 74<br>W<br>183.8 | 75<br>Re<br>186.2 | 76<br>Os<br>190.2 | 77<br>Ir<br>192.2 | 78<br>Pt<br>195.1 | 79<br>Au<br>197 | 80<br>Hg<br>200.6 | 81<br>Tl<br>204.4 | 82<br>Pb<br>207.2 | 83<br>Bi<br>209 | 84<br>Po<br>209 | 85<br>At<br>210 | 86<br>Rn<br>222 |
| 87<br>Fr<br>223 | 88<br>Ra<br>226 | 103<br>Lr<br>262.1 | 104<br>Rf<br>261.1 | 105<br>Db<br>263.1 | 106<br>Sg<br>263.1 | 107<br>Bh<br>264.1 | 108<br>Hs<br>265.1 | 109<br>Mt<br>268 | 110<br>Uun<br>269 | 111<br>Uuu<br>272 | 112<br>Uub<br>277 | 113<br>Uut | 114<br>Uuq<br>289 | 115<br>Uup | 116<br>Uuh<br>289 | 117<br>Uus | 118<br>Uuo<br>293 |

| 57<br>La<br>138.9 | 58<br>Ce<br>140.1 | 59<br>Pr<br>140.9 | 60<br>Nd<br>144.2 | 61<br>Pm<br>146.9 | 62<br>Sm<br>150.4 | 63<br>Eu<br>150.4 | 64<br>Gd<br>152 | 65<br>Tb<br>157.3 | 66<br>Dy<br>158.9 | 67<br>Ho<br>162.5 | 68<br>Er<br>167.3 | 68<br>Tm<br>168.9 | 70<br>Yb<br>173 |
|---|---|---|---|---|---|---|---|---|---|---|---|---|---|
| 89<br>Ac<br>227 | 90<br>Th<br>232 | 91<br>Pa<br>231 | 92<br>U<br>237 | 93<br>Np<br>238 | 94<br>Pu<br>244.1 | 95<br>Am<br>243.1 | 96<br>Cm<br>247.1 | 97<br>Bk<br>247.1 | 98<br>Cf<br>251.1 | 99<br>Es<br>252.1 | 100<br>Fm<br>257.1 | 101<br>Md<br>258.1 | 102<br>No<br>259.1 |

As can be seen from Table 1, silver is the best conductor of electricity. Copper is number two and aluminum is number four. Table 2 is the whole Periodic Table and depicts the location of the commonly used electrical elements like copper (Cu), aluminum (Al), silver (Ag), iron (Fe) in transformer and motor core, gold (Au) used in electronics, silicon and surrounding elements for electronics. It should be noted that Si has one of the highest boiling points among elements, its boiling point is 2357°C while for copper it is 1083°C and aluminum it is 660°C. This is why a bucket of sand is usually on standby as a fire retardant. Note the three best conductors are in the same column, with Cu, then below it is Ag and then Au. There is a great shift in the electrical industry from copper to aluminum mainly due to cost. The price of copper is raising everyday but aluminum price is decreasing every day because it is the third most abundant element on earth after oxygen and silicon. Note sand is $SiO_2$, so there are two oxygen atoms for every silicon atom. This sand atoms make up the biggest portion of the interior of the earth, the mantle and the crust. Water which fills the ocean is $H_2O$ (indicating two hydrogen atoms for every oxygen atom) is a miniscule compared to the mantle and crust which is made of sand. Aluminum, the third most abundant element is found as bauxite and with lots of electricity, electrolysis of molten bauxite produces pure aluminum.

Actually, aluminum is not at all optimum for electricity. Aluminum being number four conductor compared to copper being number two, is already a bad conductor. Plus it forms a layer of oxide (white in color) in microseconds and all oxides of metals are insulators. An energized join of two aluminum cables will therefore be arcing because electricity cannot flow easily through the aluminum oxide in-between. AndAs can be seen from the Table 1, aluminum's melting point is about half of that of copper. this

arching will lead to fires because aluminum burns at almost half the temperature of copper as can be seen from the Table 1. Therefore aluminum is a bad but very pervasive (due to price) material to use for carrying electrical power. When power companies do overhead installation with aluminum conductors, their staff are required to use torque wrench to bolt the joints very tightly to ensure that the bolt fuses the two aluminum conductors such that oxides cannot form as easily in-between. In some countries an anticorrosion liquid is poured prior to fastening the bolt.

Comparatively for copper conductors, this author has done wiring in his parents' 50-year-old home where it can be seen that the stripped copper is still perfectly shinny copper color; which means the copper wire has not corroded in 50 years compared to microseconds for aluminum. The PVC insulation helped a little in protecting the copper from the atmosphere, but the protection of the copper is not due to the PVC because this author has seen copper totally black despite stripping long lengths of PCV away from the copper wire. Copper corrosion has become an art form. Artist actually burn a plate of copper to different levels to achieve various colors. Heated copper will initially turn to golden then orange, pink, purple, dark blue, and light blue, before turning black as copper oxide but as sulphates gets in, it turns to blue and eventually bluish green as it reacts with sulfur and carbonates. The statue of Liberty of USA has reached the final color of copper. If any of these colors are observed on a copper wire, it must be filed off. **Filing seems to be the best solution** as sandpaper is hard to control and therefore will tend to sand our fingers in the process. Alternatively, a penknife can be used to scrape off the corrosion but of course a penknife is hazardous to handle near a finger. In the picture in Fig. 32, left is the original blackened wire. This was cut and stripped but the wire inside was still blackened with oxide. So, a penknife was used to scrap the black oxide away, and this motor which did not run for four years started running. It should be noted that the second color of corroded copper is orange and even at that color, a twisted join of copper cannot be soldered; solder will simply not stick to it.

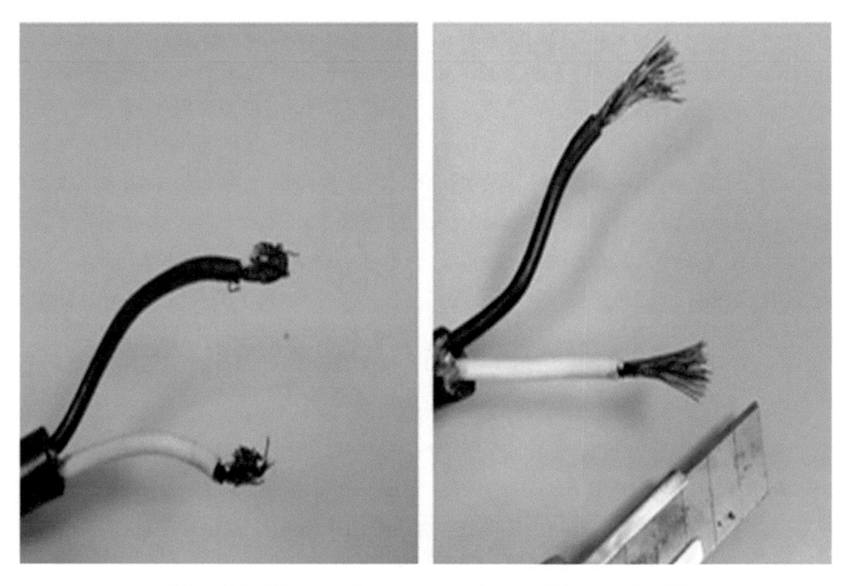

Fig. 32: Removing corrosion with penknife

So, in the electrical industry all technicians, engineers and scientists are continuously fighting the problems of oxides of copper and aluminum (corrosion). But metal oxides are used in the electrical industry in two cases as follows.

In surge-arrestors, zinc oxide (ZnO) is placed in-between the Live wires and Ground (Earth) wire which goes to the Ground (Earth) rod as shown in Fig. 33. A common term in industry for surge arrestors is SPD (Surge

Protection Device). Their function is to protect installations from lightning strikes. In a Distribution Box (DB), the surge arrestor is placed at the power incoming portion. So, in Fig. 34, it should be placed before the RCD. If lightning strikes any of the Live or Neutral wires, the bond between zinc and oxygen will break and the zinc will conduct the surge into Ground (Earth). It must be reminded to the reader that the world top electrical component companies are GE, Siemens, ABB and Schneider. This author met a technician working for a large hospital who purchased surge-arrestors costing thousands of dollars while Schneider sells it for under a hundred dollars. So, common sense must be used; how can an electrical component manufacturer ranked far below Schneider make a product and sell it for far higher cost than Schneider. Contractors need such common sense to avoid being cheated.

The other use of metal oxide is in mineral insulated cables or MI cables for short. These wires power boilers where plastic (PVC or XLPE) insulated wires cannot be used simply because the plastic will melt. In MI cables, magnesium oxide (MgO) is used to separate the three live copper live wires and N; no plastic is used. Surrounding the L and N wires is MgO and surrounding the MgO is a pipe of copper. Other than its electrical insulation properties, MgO also has a heat resistance property, thereby providing heat insulation to the copper conductor from the boiling load for example. Looking at the two compounds; ZnO has a weak bond between Zn and O which could be broken by a surge of electricity while Mg and O has a very strong bond. So, the weak bond compound is used to ensure the Zn and O separates as fast as possible and the very strong bond compound is used to ensure electrical conductivity does not happen at all.

It is sometimes difficult to remember which oxide is used in surge-arrestor and MI cables. A way to remember is when we were young, our school science teacher put a small piece of magnesium over a Bunsen burner and it burnt with a blinding bright white flame. Note the fire triangle; one side is fuel, the other heat and the third is oxygen; all three is required for fire. The intense burning with bright blinding light indicates that the fuel (magnesium) is very reactive. This burning is oxidation or adding oxygen to magnesium, or in other words the formation of corrosion. If magnesium burns so intensely, its bond with oxygen is very strong and even high voltage cannot split the magnesium ion from the oxygen ion. So, it is used as the insulator in MI cables and to give extra insulation in the highest voltage underground cables.

Zinc on the other hand will not burn or oxidize with a school science lab Bunsen burner. A very high temperature will be needed for it to burn. In fact, it is one of the least reactive metals so it is the element used to coat iron to prevent rust. The resulting iron is called galvanized iron or GI for short. So, the weak bond in-between zinc and oxygen is utilized in surge-arrestors as shown in Fig. 33. The surge arrestor is the incoming protective device of newer home and offices. Most home has the incoming device as the RCD as shown in Fig. 34. In some countries the incoming device is the Isolator. The other protective device is the MCB (Miniature Circuit Breaker). So, of the three protective devices only the isolator is a dumb device; it is a manually operated device without intelligence and is used to manually switch off the power to enable performing electrical work. RCD trips if there is a Ground (Earth) fault and MCB trips due to too much current being drawn (overload) or a short circuit. This new electrical component called surge-arrestor should be placed at the incoming device before the RCD or Isolator. In a single-phase home, L and N go into the top of the device and then is jumped to the RCD. At the bottom of the surge-arrestor is a green wire going to the Ground (Earth) bar and thenceforth to the Ground (Earth) rod. In normal circumstances, the zinc oxide acts as an insulator in-between the incoming L and the Ground (Earth) rod. But when lightning strikes, the Zn easily dissociates from the O and lightning surge travels through the Zn to the Ground (Earth) rod; thereby protecting the home.

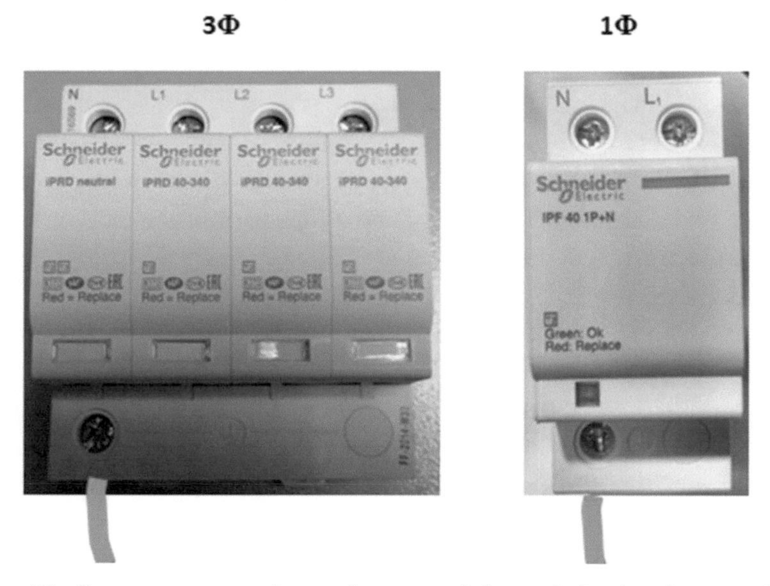


Fig. 33: Surge arrestor, three-phase on right and single-phase on left

Fig. 34: Three phase Distribution Board

Soldering is the best way to join wires. This author does not like crimping unless in some projects where it is a requirement by the customer. Crimping stresses the crimp lugs and the cable within. In mechanical engineering, it is a standard practice to burn the metal after stressing it. For example, in making a car bonnet, the sheet of steel is first clamped and then heated to relieve the stress. Crimp lugs are not heated after stressing it with a crimping tool so, stress remains in the lug and wire causing fast corrosion and deterioration of the join. On the other hand, in soldering, two wires are twisted (stressed) then soldered which is a heat treatment to relieve stress plus enable a very strong joint. If the two sides of the join are pulled hard enough, the wire will not break at the join but at the copper portion; such is the strength of the join. It must be noted that in making joins of copper wires, lots of twists must be made to ensure as long as possible length of copper to copper connection. The solder is applied outside, mainly for mechanical strength. If the copper-copper connection is good enough due to the many rounds of twisting, the current will avoid flowing through the solder (electricity always takes the easiest path). Of course, the join should not be twisted too much, it is a skill to be able to realize how many twists a wire can withstand without breaking, but heating it later with solder will relieve much of the stress. Solder should never be allowed to get in-between the copper. That is, never have copper-solder-copper join. This is because solder is 60% tin and 40% lead. As can be seen in Table 1, tin is number 10 down the list in conductivity while lead is way down. Therefore, solder is a bad conductor of electricity so a copper-solder-copper join will be a heat point and eventually an electrical fire point. Solder is a good mechanical fastener. If a soldered join is pulled, the soldered point will not break as fast as the solid copper potion. So, the lots of twisting are not for mechanical strength but to ensure as big a surface area as possible of copper-copper connection. Soldering can be done with a gas soldering iron. If an electric soldering iron is used to solder a live wire, it will trip the RCD. If a gas soldering iron is not available or if your gas container is empty, the last resort is to use a candle, it is fast and effective except that there is lots of carbon formed in the process, this carbon can later be wiped out. So always keep a candle in a good container in your toolbox, this author keeps it in the plastic that comes with a big drill bit. Without a container, it will mess up your toolbox. Be careful when soldering a live wire, many people forget the solder lead they are holding is also conductive.

This author came up with a fast and effective method of soldering for construction. The correct method of joining two copper cables is to strip both ends a little longer, about two inches, bend both joining portion 90 degrees from the orientation of the wire. Then a combination plier is used to twist the whole two inches of wires tightly. It does not matter if the copper is stressed. Then long piece of soldering lead (about one feet) is taken out from the roll of lead, it is very much harder if the whole roll of solder is held in the hands. Then a butane flamer, which is commonly found in a hardware stores is lit up and blown onto the end portion of the twisted joint while the soldering lead is gradually lowered from top onto the joint. Heat will quickly conduct from the end of the twist to the beginning of the twist and so the soldering lead is slowly moved towards the beginning of the twist. If the soldering lead is lowered to the flamed portion of the join; the solder will not stick to the copper at that portion because the copper there will already have a layer of orange copper oxide. Heated copper will initially turn to golden then orange, pink, purple, dark blue, and light blue, before turning black. These colors represent different concentration of copper oxide on the cable. Soldering lead do not stick to copper oxide. Which is why the soldering lead must be dropped on the non-flamed portion. The flaming also relieves the stress which was exerted onto the copper atoms by previous hard-twisting of the copper with the combination plier. The hard-twisting must be done to ensure the solder lead do not go in-between the copper. Then wait a while for the joint to cool then bend it towards the cable and tape it with 3M tape. The bigger roll (about 3' diameter) of 3M tape is better than the smaller roll (2" diameter).

In the experience of this author who has a technical school, there is a three-phase induction motor in the school as shown in Fig. 35 below:

Fig. 35: Autotrans motor circuit with motor terminal connection without crimp lugs

The termination at the six motor terminals used to be crimped. But in a technical school where the motor incoming wires are constantly moved many times a day; the crimped join to the motor will last a maximum of two days before it breaks. One day, a student showed a way to connect without crimping as shown in Fig. 35. A wire stripper was used to move up the PVC about 1.5 cm, the strands of copper separated about equally (left and right having about equal number of strands) and through this hole, the motor termination screw is moved in and the nut is turned on the screw. There is very little stress to the wire, no soldering is needed. Once this author adopted this method, plus tying the incoming wire with a cable tie via a screw hole, the motor termination of all six wires did not break for four years. Meaning the crimped joint lasted for maximum of two days and this method lasted four years.

Fig. 36: Screwing on PVC insulation

Another common problem in wiring is screwing onto plastic insulation as shown in Fig. 36. This author believes this is caused by many technical schools instructing students to ensure the copper wire should not be seen after wires are screwed to a switch for example. In fact, two percent marks are deducted for every time copper can be seen. The typical examination is wiring of a home so there are many terminations and if copper can be seen at all terminations a student can fail the practical examination just because he allowed copper to be seen. So, the students strip as little as possible and therefore have a higher chance of screwing onto the PVC (polyvinyl chloride) or XLPE (cross linked polyvinyl chloride) insulation. One 20-year veteran of the electric industry said this is the most common problem he has to repair all over the state. Equipment will work initially as there is a little spot of copper conduction at the edge of the PVC. But this spot is of course not sufficient to carry the designed current so it will heat up as the atoms within it vibrate not allowing the electrons to flow easily. This heating up will cause the atoms to vibrate even more causing even more resistance to electron flow. Eventually if a serious fire does not break out, small arcs will form oxides which will eventually render the conductivity through this join to be zero and the equipment will fail to work.

This author repaired a MSB which was built perfectly. Perfectly meaning, all the designs are sound with quality electrical components. Also, all screws are tightened to very tightly which should be the case. The best way to achieve this is with a long, probably 1 m long screwdriver. This author actually wished he had a 1 m long screwdriver to open many screws while repairing the MSB but had to be content with a one feet screwdriver and a whole lot of strength. But it was later detected that this whole MSB would not work to protect the factory. The MSB has intelligent O/C and E/F relay giving fault signals via the Shunt-Trip cable to the main MCCB to trip it. But in this case, there was a fuse on the L wire before the main MCCB. This author will not recommend a fuse here. This critical Shunt-Trip wire is sending fault signals to the MCCB via this wire, in case the fault signal burns this fuse, the MCCB will never trip. Worst of all, in this particular case, the Shunt-Trip L wire at the fuse was screwed on to the PVC portion, so no signal can ever trip the MCCB. Meaning a serious short-circuit in this factory will never cause the MCCB to trip and therefore it will cause thousands of amperes to flow in the wires without stop, thereby burning the wires and therefore causing electrical fires that are known to engulf whole factories and buildings.

The resistance R of a conductor of uniform cross section can be computed as:

$$R = \frac{\rho L}{A} \qquad (18)$$

where

L is the length of the conductor, measured in meters
A is the cross-sectional area, measured in meters square
ρ Greek letter rho is the electrical resistivity (also called specific electrical resistance) of the material, measured in Ohm · meter. Resistivity is a measure of the material's ability to oppose electric current.

Some people this author met asked won't the fuse prevent damage from a short circuit. Actually, most fuse are designed to protect for overload conditions; meaning if a motor is supposed to draw a maximum of 4A, a 6A fuse is used. If there is a bearing jam, the motor will draw more than 4A to overcome the jam and if it draws more than 6A, the fuse will break. A short circuit is different. If L and N touches at the wire going to the motor, it is a short circuit. Using the formula below:

$$V = IR \qquad (19)$$

$$\frac{V}{R} = I$$

$$\frac{240V}{0.03\Omega} = I$$

$$8000A = I$$

So, the short circuit of L to N resulted in a current or 8000A while the normal operating current of this motor is 4A. If a protective fuse of MCB do not trip in time, 8000A will do some serious damage; most probably a fire. The cables will burn first, then the whole building. This is the reason all MCB has two ratings. For example, the lowest home MCB may be rated 6A which is the overload rating. There is another rating on the MCB like the number 6000 in a box, this is the short circuit rating. The 6000 means this MCB can trip even if a short circuit of 6000 A flows through the MCB. But the standard household MCB is a mechanical device and though mechanical speeds seems fast for human beings it is extremely slow compared to the almost-the-speed-of-light speed of electricity. If the 8000A flows to the motor due to a short circuit, even if the MCB does trip, lots of damage could be to the equipment especially if the equipment being run is an electronic one, like an inverter-controlled motor. The standard house-hold fuse is also slow to blow to prevent surge from going into the equipment. This author has seen $100,000 equipment being damaged by a short circuit in a ballast of a florescent tube. So, the standard practice in industry is to protect expensive equipment with a HRC (High Rupturing Capacity) fuse. High in HRC stands for high current as from a short circuit or lightning surges. The top of the line HRC fuse is made by Bussmann. Most big industrial companies trust Bussmann. A Bussmann fuse the size of the one in a standard plug-top which cost $0.05 can cost $60. But in this author's experience it is better to source it directly from Bussmann because the profit margin a cheater can make by purchasing a $0.05 fuse, coloring it with Bussmann symbols is so high, someone may already have a business doing it.

Another part an electrician must watch out for is carbon brush. An original carbon brush for a Hitachi or De Walt drill may cost $20 while the cheapest carbon brush can cost $0.25. There is no way an electrician can determine if the $20 carbon brush they bought is actually the real thing unless they source it from the original supplier (Hitachi or De Walt). This author got cheated once while working in the Western Digital factory. A super long-lasting carbon brush (more than 10 years of continuous operation) from Bodine Motors was faulty but a local supplier supplied the $0.25 carbon brush with all the marking of Bodine Motor carbon brush.

# Chapter 9

# Motors

## 9.0 Induction motors

Motors consume 65% of industrial power use. The most common motor used is the induction motor which is an ingenious invention of Nikola Tesla. In a typical large generator, a high current is initially generated as the DC magnetized rotor passes the L1 (R) coil then it passes the L2 (Y) coil and then in the L3 (B) coil respectively. In the faraway (maybe 1000 km away) induction motor, located in a population center, a magnetic N is generated first in the L1(R), then in the L2 (Y) and then in the L3 (B) coils. The induction motor stator coils thus get magnetized almost as soon as the respective coil in the 1000 km generator coil gets magnetized. As shown previously in equation (11), in a generator, as the magnetic rotor passes a stator coil, it will energize the respective stator coil in a 1000km away induction motor coil in 0.003 seconds. In effect the magnetization of the induction motor stator coils follows the rotation of the generator rotor. So, the induction of current in the generator coils created a whirlwind of magnetism as Tesla termed it, around the stator coils of an induction motor, located say some 1000km away. This whirlwind or consecutive magnetization of the stator coils of an induction motor is today termed the Rotating Magnetic Field (RMF). Actually, this RMF in the induction motor is not moving at the same speed as all generator rotors but a certain multiple of this following equation (22), except for steam turbines which turns at the grid frequency. As seen from the calculation below, it only takes 0.003 seconds for induced current in the stator coil of a generator to magnetize the corresponding coil in an induction motor 1000km away.

$$\mathrm{S} = \frac{D}{t} \quad (20)$$

Where S= speed, D=distance, t=time, so

$$t = \frac{D}{S} \text{ or } \quad (21)$$

$$t = \frac{1000\,KM\;X\,1000m}{3\,X\,10^{8}} = 3.3\,X\,10^{-3} = \underline{\underline{0.00333\,seconds}}$$

The calculation assumes electron flow is almost the speed of light which is true in bare overhead lines but as insulators are placed over the conductors, the speed of flow of electrons will reduce significantly. This reduction of speed is caused as heat builds up around the enclosed conductor. Also, underground cables is shaped like a longitudinal capacitor which is energized two times every cycle.

The rotor of the induction motor will try to keep-up with the rotation speed of the RMF but it doesn't keep up and has a speed lag called the slip. In synchronous (same speed) motors there is no slip, meaning the rotor turns at the same speed as the RMF. Most work done in industry is done with induction motors, mainly because it requires no carbon brush to send electricity to the rotor. Synchronous motor needs a carbon brush to send DC to the rotor. Therefore, induction motors can be made to last up to sixty years, as long as the varnish around the copper coil and the two bearings that hold both ends of the rotor are of good quality. The most modern high-speed induction motor uses Active Magnetic Bearing (or AMB). Here the shaft is levitated in space by actively controlled electromagnets leaving zero contact in-between stator and rotor. These are called Ultra-High-Speed Magnetic Bearings. With no friction, magnetic bearings do not only have a huge increase in lifetime but also enable vacuum operation with very high rotational speeds. Rotation speed of up to 500,000 RPM (Revolution Per Minute) have been achieved.

In the first paragraph, the details of the RMF was discussed but putting it simply, in the stator of an induction motor, the three-phase coils are arranged as in the generator of a power station. Thus, the RMF is created in the stator of the induction motor that rotates synchronously with the rotor of the far away generator. The rotor then tries to catch up with this RMF but does not catch up, it only achieves 95% the speed of the RMF.

Looking at one particular bar of the squirrel cage rotor with respect to the RMF in the stator. In slow motion, at the start, the RMF is switched on as someone switches on the three-phase induction motor. The RMF field lines will cut the stationary bar of the squirrel cage rotor. By the right-hand Fleming's rule, a current will be induced in this bar. This current will flow to the end rings, go down a semi-circle and then flow through the bar right below the first bar up the second end ring, flow a semicircle in this end ring also and back to the first bar. Therefore, at one moment of time, the electrons are moving in a square circuit. This current will produce a magnetic field using the right-hand thumb rule. This magnetic field will lock with the rotating magnetic field (RMF) and the rotor will start turning, to eventually approach the same speed as the RMF (synchronous speed). As the rotor increase speed with respect to the RMF, it experiences even more varying-Field-density of magnetic lines as it cuts even more field lines per time period. It thereby gains even more current flow in the bars and therefore increasing magnetic flux lines around itself. The bar will eventually catch up with the RMF (achieve synchronism). But at that moment, the bar is not experiencing a varying-**F**ield-density (F out of the FBI required to generate current flow), since it is moving in synchronism with the RMF. It is just like two cars travelling at the same speed, the passengers of both cars can talk to each other because the relative speed between one car and the other is zero. Since the bar is not experiencing a varying-magnetic-**F**ield-density, the current flow in the bar will reduce to zero and it loses its own magnetic field lines. Without the magnetic field lines, the 'string' of magnetic force to hold onto the RMF is gone so the bar will fall back from the RMF. But, as it falls back, it is again experiencing a varying-**F**ield-density and thereby gains current flow and consequently its own field lines. With regained 'strings' of field lines to hook onto the RMF, the bar will again try to catch-up with the RMF. This continuous catching up and falling back all throughout the rotation of the bar around the motor stator, in-effect causes the rotor to be moving slower than the RMF by about 5%. This 5% is called the slip. The synchronous speed in RPM (Revolution Per Minute) is given by:

$$\eta_s = 120\frac{f}{p} \quad (22)$$

Thus, an induction motor moves at:

$$\eta_s = (120\frac{f}{p})\text{ X }95\% \quad (22)$$

Synchronous motors turn at synchronous speed which means that rotor rotates the same speed as the RMF in its stator. And this RMF speed in the stator of a synchronous motor is the same speed as the RMF in the stator of a steam turbine generator located say 1000km away. This is because the steam turbine's RPM is the same as grid frequency. Note that this is only the case for steam turbine generator; for hydro or gas turbines, the RPM of the generator rotors are multiples of this speed following formula (22). An induction on the other hand turns at 5% less than that speed. But the problem with synchronous motor is that it needs a carbon brush to send DC into the rotor to turn it into a magnet. Starting in 2015, synchronous motors have been used to move robot arms. This is because humans have achieved perfect control of the RMF around the stator of the synchronous motor. This is enabled with the development of the cell phone software called DSP

(Digital Signal Processing). Previously this author went to the USA to study telecommunication engineering because his dad and brother worked for the main telecommunication company in the country. But in those days, signals are translated to its equation so as to manage them. This involved very heavy-duty calculations, double-integration, triple-integration and circular-integration and so forth. Student taking such courses can achieve only grade Cs and Ds. So, this author did not pursue that study. But later DSP came along which simply coverts the complicated waves into quantified heights at different periods of time. The shorter the span of time on the X-axis that the signal height is counted (called the sampling rate), the more accurately the signal is defined. In sound systems, higher sampling rate gives more melodious songs. So, a software is used to code all this, called DSP which enabled the evolution of cell-phones. By the year 2015, DSP was used to perfectly control RMF around synchronous motors such that it can be used to move robot arms. Previously robot arms were moved with stepper motors. Stepper motors are described further down in this chapter. But stepper motors are known to be not so powerful. With synchronous motors, even robots as strong as the 'Transformer' movie ones are possible as long as there is a nuclear power source (or some power source like that). Induction motor cannot be used for this because of the lag of 5%. If the motor lags is a consistent 5% it would have been alright but the speed is not a consistent 5% less than synchronous speed. At different moments it can be 99%, 98%, 97%, 96% or 95% of the RMF speed. Robots cannot be used if there is such variance. In the Western Digital factory where this author worked in, the robot picks up computer hard disk platters and place them in cassettes which has 0.5 mm space beyond the disk thickness. And, this movement is happening so fast that the eye can barely keep up with the movement. So, if the robot is slightly not accurate, all the computer hard disks will be damaged. But with synchronous motors, where the rotors follow exactly the RMF, it can be used to operate robotic arms.

The P in the equation is the number of poles of the motor. For the two-pole delta connected induction motor shown in Fig. 37, each phase is joined to a coil and the other end of the coil is joined to another phase. Thus, each phase forms only one coil which is a solenoid having one N and one S, that is how it is called a two-pole motor.

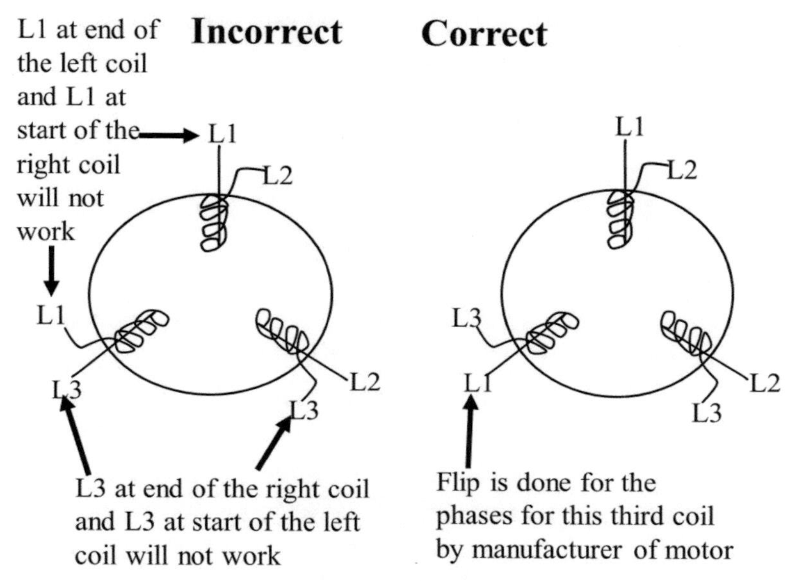


Fig. 37: Two-pole induction motor connection

Fig. 27 was drawn with excel and is the waveform for the two-pole three-phase induction motor shown in Fig. 37. Going by L1, L2, L3 in sequence, the 1st coil should be L1, L2 then L2,L3 then L3,L1. But as explained in Fig. 37, this will not work, the phases need to be flipped on the third coil so the wires going into the three coils from top are, L1,L2 then L2,L3 then L1,L3. This is because it won't work for the outgoing of the second coil to be L3 and the incoming of the third coil to also be L3. Also, the outgoing of the third coil cannot be L1 and the incoming of the first coil to also be L1. But installers of induction motors do not need to bother about this as this flipping is already done by the manufacturers. So, there are six connection terminals of a three-phase induction motor, three up and three at the bottom. Installers just have to connect L1,L2,L3 on top and L1,L2,L3 at the bottom to get it running.

The **line voltage** (phase to phase voltage) in the three motor coils are the purple area between the phases as shown in Fig. 27. The dotted lines indicate the three coils being energized sequentially around a three phase two pole induction motor. Four-pole induction motor is the most common but two-pole is easier to understand, the same logic is just expanded in a four-pole induction motor. It can be seen that coil 1 has 0V at 30°, coil 2 has 0V at 90° and coil 3 has 0V at 150°. So, there are 90-30 = 60° between coil 1 and coil 2 being 0V and also 150-90 =60° between coil 2 and coil 3 being 0V.

For a 60Hz American system, the motor turns at 60 cycles---1 sec, =>1cycle---1/60=0.0166=16.6ms for a full cycle. It takes 16.6ms for a full cycle in 60 Hz system, so, 60°---16.6ms => 240°---16.6/360 X 60 =2.766ms between one coil of an induction motor deenergizing to the second coil deenergizing.

For a 50Hz British system, the motor turns at 50 cycles--- 1 sec, => 1 cycle---1/50=0.02=20ms. It takes 20ms for a full cycle in 50Hz system, so, 360°---20ms => 60°---20/360 X 60 =3.33ms between one coil of an induction motor deenergizing to the second coil deenergizing.

The most common induction motor is the four-pole motor. In this motor, each phase is connected as shown in Fig. 38; each phase forming two coils or two solenoids. With two solenoids there will be a N and S for one solenoid and another N and S for the other solenoid, that is four poles. Thus, the number of N and S poles formed by each phase wire is how the number of poles of a motor is categorized. The four-pole motor speed is 1800 RPM (1500 RPM). This is derived as 120 X 60/4 = 1800RPM (120X50/4=1500 British). In a standard four-pole motor, each two phases will form two coils so there are six coils.

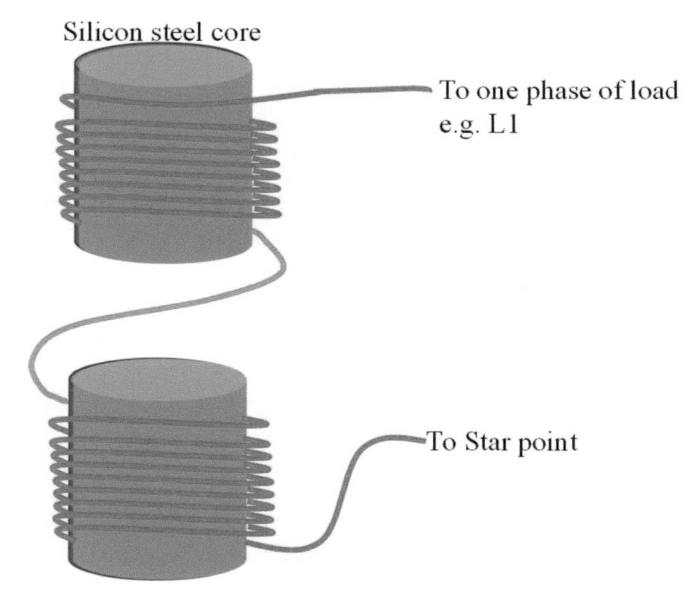


Fig. 38: Winding method

If each phase wire forms a clockwise coil and the other end continues to form an anticlockwise coil, four poles are formed; N and S on the clockwise coil and S and N on the anticlockwise coil. This is a four-pole motor which will spin at 1800 RPM (using 60 Hz) and 1500 RPM (using 50 Hz). This is the most common motor in use.

Now if we look at the squirrel cage, when induced current is maximum on one particular bar, it will be of reducing strength in the nearby bars but on all these nearby bars, current will flow in one direction (say from left to right). This induced current flows to the end rings (left to right in Fig. 39) and then flows down the right end ring, continue travelling in the bottom squirrel cage bars from right to left, reach the left end ring and so forth. Therefore, a circulating current is formed which flows from left to right in the top bars, down to the bottom the left end ring, continue flowing from right to left at the bottom bars, into the left end ring and back up to the top of this end ring and repeat itself. This circulating current creates magnetic field lines that interact with the stator RMF produced by the generating station.

Fig. 39: Squirrel cage of an induction motor

Induction motors had a problem in that it spins at a fixed speed or RPM which is synchronous with the generator in a steam turbine power station minus 5%. This hampered its wide use previously. It was only after the invention of the IGBT (Insulated Gate Bipolar Transistor) in 1982 that the use of induction motor started becoming widespread. With an IGBT, induction motors need not follow the generator rotation. RMF can be created with the help of a computer plus an IGBT. So, an induction motor can be controlled precisely from a speed of a few RPM to much higher speed than synchronous speed. The fastest speed of an induction motor if it is just connected to the grid is if it has four poles is below at 60Hz.

$$\eta_s = (120\frac{f}{p})\ X\ 95\% \qquad (23)$$

$$\eta_s = (120\frac{60}{4})\ X\ 0.95$$

$$\eta_s = 1710\ RPM$$

at 50 Hz:

$$\eta_s = (120\frac{f}{p})\ X\ 95\% \qquad (24)$$

$$\eta_s = (120\frac{50}{4})\ X\ 0.95$$

$$\eta_s = 1425\ RPM$$

In an induction motor, the rotor is made of bars of conductors (aluminum or copper) arranged in a circle and held together by conducting rings at both ends. In effect it is shaped like a squirrel cage, thus the name squirrel cage induction motor. The squirrel cage bars are arranged slightly skewed and not parallel to the shaft to:

1) reduce humming noise
2) to avoid stalling the motor

Squirrel cage bars and end rings, made of aluminum will have low mass and therefore a lower centrifugal force is needed to turn it, thereby making it easier to start. But if squirrel cage bars and end rings are made of copper, it can start with higher loads and have higher overall efficiency and thereby can reduce of $CO_2$ emission. Within the squirrel cage is stacked laminated steel sheets of steel for the following reasons:

1) Steel (about 96% iron) will allow much better flow of magnetic field lines (better permeability) than free air. Iron has the highest permeability among elements. In fact, permeability of air is 1 and so is the permeability of most elements like silver, copper, aluminum or gold but the permeability of iron

can reach 100 and with some addition of other elements it can reach even reach 10,000. Thus, with iron laminates, the magnetic field lines around the squirrel cage bars can made very much stronger as current is induced and flow within them.

2) A squirrel cage structure by itself is not capable of carrying a big load. Imagine an empty squirrel cage structure carrying a load requiring 300 HP for example; even its look doesn't depict a lot of mechanical strength. The laminates of steel provide mechanical strength to the squirrel cage structure.
3) Laminates of steel with varnish in-between will prevent the formation of huge eddy currents (unwanted storm of electron flow) which can affect the designed interaction of the circulating stator field lines with the squirrel cage bar's field lines of the rotor. This eddy current, if unchecked can also produce lots of heat in the rotor. Eddy is still formed in the outside body of the induction motor but steel laminates in the stator (around which the stator windings are wound, enable keeping the main steel body of the induction motor a distance away; thereby reducing eddy current in the outside body. This author has tried running a 25 HP three-phase motor with only the three phase wires going in and removing the green Ground (Earth) wire. The motor turns but was very jerky making a "de de de de" sound and is physically vibrating. By just connecting the Ground (Earth) wire, the sound changed to, "deeeeeeeee" and the vibration stopped and it is perfectly smooth. So, the eddy current in the body of the motor, if not leaked to Ground (Earth) via the Ground (Earth) wire will have a drastic affect to the functioning of the induction motor. This means, in factories with lots of induction motors Ground (Earth) is quite critical. In the factory where this author worked, at one time the whole factory tripped every day at about 6pm. Electricians were called in but the root cause cannot be found for some time. So, the contractors planted a whole lot of Ground (Earth) rods in the land behind the factory and the problem disappeared. So that factory has a very good Ground (Earth) now. Thinking back this author has an idea of the root cause of the tripping. This is a hard disk platter manufacturing factory. The activity that occurs at around 6pm is visual inspectors (exclusively ladies) taking a sample of the production disks to inspect for defects. They use a table lamp which is of course of high quality made of iron. In high tech factories we cannot use standard plastic table lamps, customer who come for site visits will be nervous about buying our products if they see household lamps. In the metal pipe of the lamp is the three-core wire going to the lamp. As the lamp is moved, eventually sharp edges will cut the three-core wire causing a L-G (L-E) or L-N fault. These ladies will call technicians who are exclusive male and they will cut the damaged point, join the three wires at the same point and insulate all three wires with insulation tape. Then they put insulation tape over all three wires and it looks beautiful enough to impress the lady visual inspectors. This author has done exactly same thing early in his life and have seen a short occur after a few years. The insulation tape simply conducts after a few years. Of course, this will not happen if 3M insulation tape is used. Therefore, if a normal insulation

tape is used, keep the three joins separated by air. This looks ugly but is the best for preventing short circuits. The tape is only to prevent rats in ceilings or humans from getting an electric shock.

4) If the laminates of steel are replaced by solid steel, the magnetizing of only the squirrel cage bars which is needed for the motor to run will not happen efficiently. The whole solid steel will also have induced magnetism running around in disarray.

## 9.1 Motor Starters

The essential functions of motor Starters are to prevent a sudden inrush of voltage and therefore current into the motor coils. The sudden inrush will cause mechanical stress on the bearings and electrical stress on the motor coils. Described in this section are some of the most common methods that are used to start an AC induction motor. These methods offer some benefits in terms of cost.

## 9.2 Reduce-Voltage Starting

When starting voltage must be lowered. The common types of reduced-voltage starters are as follows. Going down the list from 1-4 one can observe the motor starting smoother and smoother. Using 1 will start a 10HP motor but the motor will jump a little. As method 4 is used, the motor starts without vibrating even a bit.

1) DOL (Direct-On-Line)
2) Star-delta
3) Auto-Trans
4) Primary-resistance
5) Soft starter using SCR (Silicon Control Rectifier) – mostly used with DC motors.
6) VFD (Variable Frequency Drive) or VSD (Variable Speed Drive) or Inverters using IGBT

## 9.3 Reason for Motor Starters

By reducing the voltage to about half prevents coils of the motor from burning because of the high ampere drawn initially times a high voltage is too much power for the coils to handle.
Example for the star-delta circuit:

Initially: P = 240V X (High amperes)
Later: P= 415V X (Low amperes)

## 9.4 Direct-On-Line starters

Though this is the simplest starter, it is by far the most common method to start a motor. It basically utilizes a three-phase contactor to send all three-phases to the motor. Why a contactor is used instead of a human hand operated switch is because the human hand is too slow. This is especially the case if the motor is very large. So, a human finger presses a start button and the contactor NO contact is used to hold this signal which sends current to the contactor coil. The contactor closes very fast energizing the three phase wires going into the motor terminals. On the contactor is written L1, L2, L3 as the incoming phase cables and on the other side are U1, V1, W1 which are the names of the motor in-coming terminals. Because of the improvement in the technology of varnish and bearings today, up to 40 kw motors are today using DOL starters. Meaning the in-rush of current and mechanical jerking can be handled by the improved varnish and bearings.

## 9.5 Star-Delta Starter

Prior to understanding how the Star-Delta motor starter works the electrical contactor must be understood. The Schematic in Fig. 40 is that of the contactor. The exact mechanical arrangement or position of electrical contacts will vary among different brands but the concept of how contactor works is shown in Fig. 40. If there is no electricity in the solenoid, the NC (Normally Closed) contacts are closed. The word Normally means when no electricity is sent in. If electricity is sent to the solenoid, it becomes a magnet and pulls the top iron bar which separates the NC contacts. At the same time, it pulls the bottom iron causing the NO contacts to close.

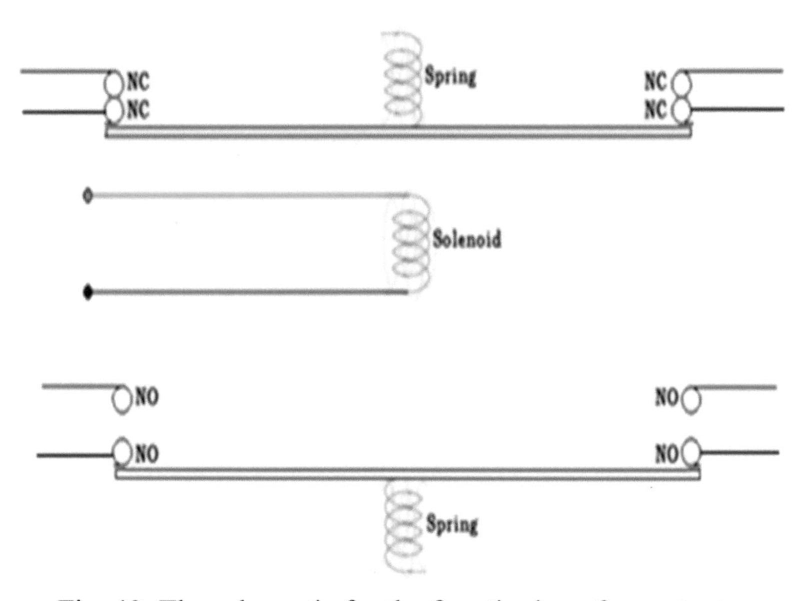


Fig. 40: The schematic for the functioning of a contactor

The power circuit schematic of the Star-Delta is shown in Fig. 41. The contactors in the Star-Delta circuit are named Line (L), Delta (D) and Star (S). Note the contactors are the same but naming them differently means the wiring to them must be changed accordingly. Initially L and S close and D is open. Thus, looking at the U1, U2 coil of the motor, one side will get 120V (240V British) from R phase (red wire or L1) and the other side goes all the way to the S contactor where R, Y, B (Red wire, Yellow wire and Blue wire or L1, L2 and L3) is looped providing 0V. This will give 120V (240V British) across the motor coil U1, U2. After 6s, the S contactor opens and D closes continuing to give 120V (240V British) from R phase one side of the motor coil U1, U2 and 120V (240V British) from Y phase on the other side of the same motor coil. This will provide $120\sqrt{3}$ =208V ($240\sqrt{3}$ =415V British) across U1, U2 coil. The other two motor coils (V1, V2 and W1, W2) will also first experience 120V (240V British) then 208V (415V British) energized by the other phases.

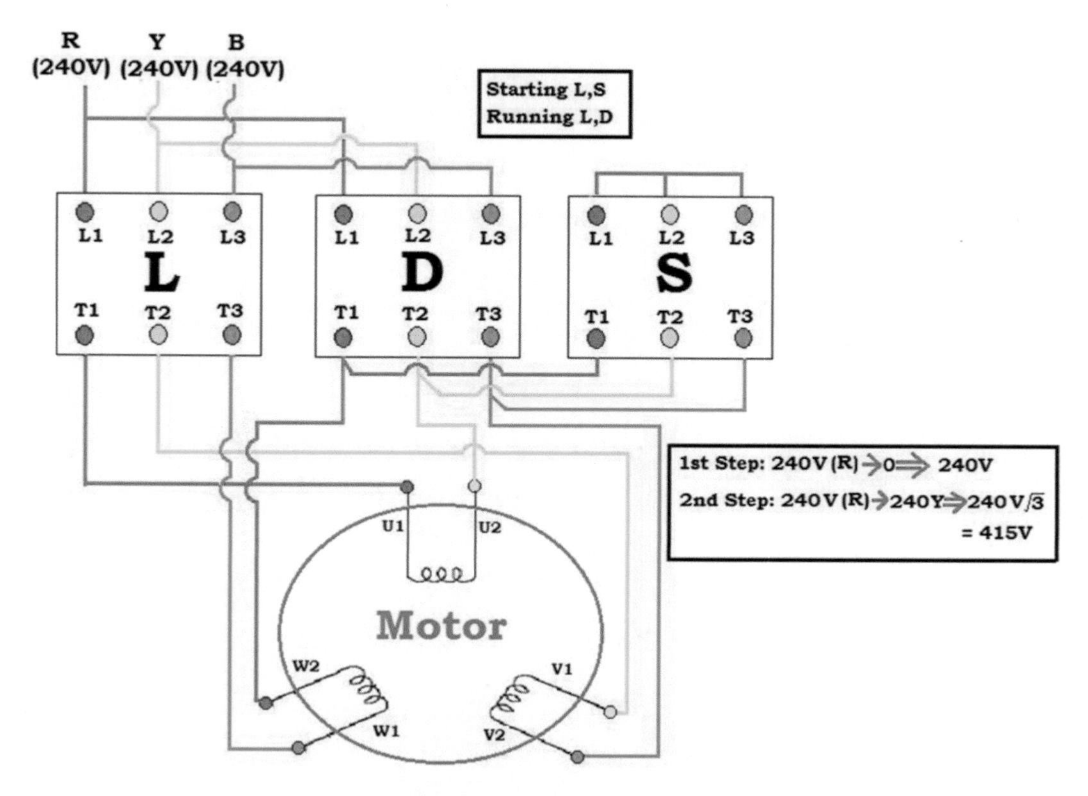


Fig. 41: Star-Delta motor starter power circuit

Fig. 42 below is the control circuit to achieve this. Initially the Start button is pressed allowing energy to go all the way to the coil (solenoid) of S contactor. This will cause the coil to be magnetized closing S1 contact. This will give energy to the three loops whose ends are the coils L, t and G. G is the Green bulb. L coil energized will result in L1 being closed which is a holding circuit for L coil. L2 will hold the push button on which a human finger presses and then releases in a portion of a second. t energized means the central contact in the timer will switch from the NC t2 to open position in six seconds, this will de-energizing S coil. At the same time the NO t1 closes energizing D coil. This will close D1 contact which is a self-holding circuit for D coil. Note the t1 switching is only for a short time similar to the human finger pushing a push button. So, a holding circuit is needed to hold the t1 switching. The D1 contact is thus a self-holding for the D coil. The D2, D3 and S2 acts as an interlocking system to prevent wrong sequence of operation of the contactors. This is necessary because sometimes this same circuit is used to operate up to 500 HP motors and wrong operation can be a major disaster.

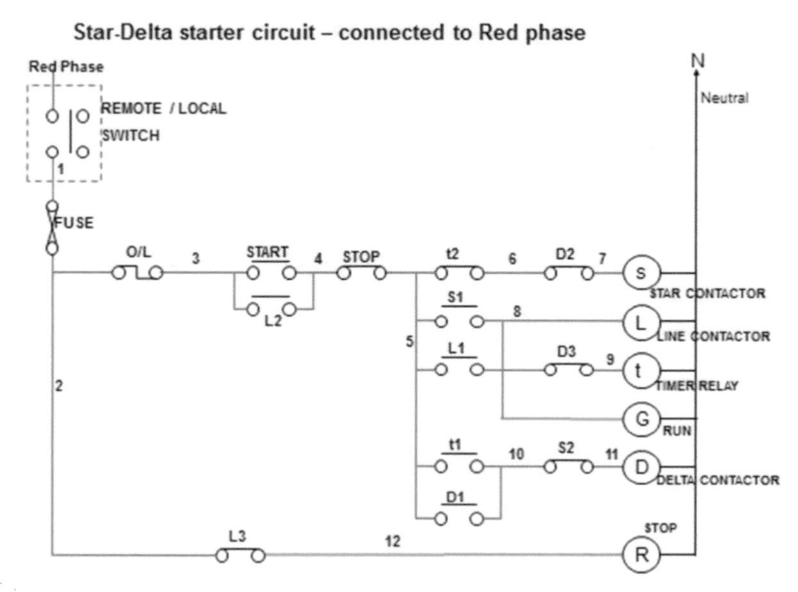


Fig. 42: Star-Delta motor starter control circuit

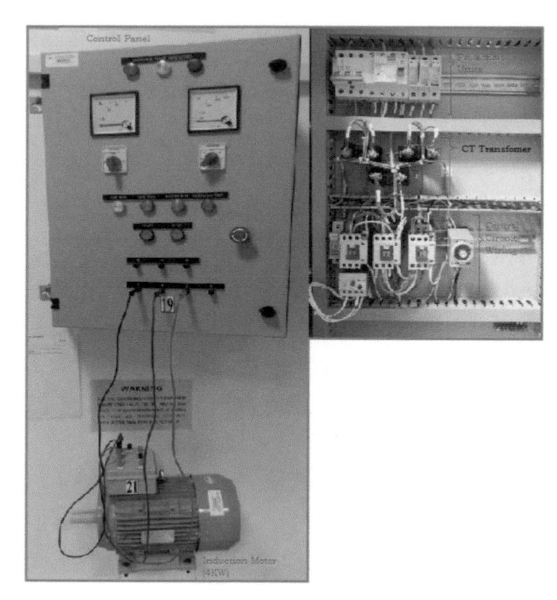


Fig. 43: The Control and Power combined circuit of a Star-Delta motor starter

## 9.6 Autotrans Starter

For the Autotrans circuit in Fig. 44, the names of the three same contactors are changes to L (Line), T (Transformer) and S (Star). Note the contactors are the same but naming them differently means the wiring to them must be changed accordingly. The circuit can be understood by looking at one Auto-trans and one motor coil. There are three separate Auto-trans sharing one core. Each of these are designed to split each of the three phase voltages L1, L2, L3 (R, Y, B) to 70% initially. That is 70/100 X 120 = 84V (70/100 X 240 =168V). In the second stage (after 6s) the Autotrans is not utilized; the incoming phase wires are directly connected to the motor coils a configuration known as Direct On Line (DOL).

To understand this one Auto-trans and one motor coil is studied. Initially T and S are closed, this will energize the Autotrans. The T will provide L1 (R) phase 240V to the top of the Autotrans coil and the S (above which L1, L2, L3 (R, Y, B) are looped) will provide 0V at the bottom of the Autotrans coil. Thus, the wire joint to the 70% point of the coil will provide 70/100 X 120 =84V (70/100 X 240 =168V British). This 84V (168V British) from L1 (R) phase will go to the one side of a motor coil and another 84V (168V British) from Y phase will come from the second Autotrans. With 84V L1 (168V R British) phase on one side of the motor coil and 84V L2 (168V Y British) phase on the other side of the same motor coil will result on $84\sqrt{3}$=145V ($168\sqrt{3}$ = 290V British) across the motor coil.

After 6s, L and S are opened basically taking the Auto-trans from the circuit. And L is closed resulting in 120 V L1 (240V R) phase on one side of the same motor coil and 120V L2 (240Y) phase on the other side of the same coil. This will result in a voltage of $120\sqrt{3}$ = 208V ($240\sqrt{3}$ = 415V) across this motor coil. The control circuit to achieve this is in Fig. 45.

So, this motor coil initially experienced 145V (290V British) and after 6s it will experience 208V (415V British). This is compared with Star-Delta where the motor coil will first experience 120V (240V British) and after 6s it will experience 208 (415V British). But Autotrans motor controller is smoother for the motor. A 10HP motor can visibly be seen to jump a little with Star-Delta motor starter but not with Autotrans. Primary resistance starters are even more smooth than Autotrans starters. And as a VFD is used it will be even more smooth. But more often than not a VFD is an “over-kill” because the price can be 40 times higher or 4000% higher. This author has seen a university where the air-conditioning system is driven by a large VFD but because it has spoilt, the students have not air-conditioning. An autotran could have solved the problem. Besides motors are improving and increasingly they are able to handle surges of current during startup and the bearings can handle the sudden jerks during startup.

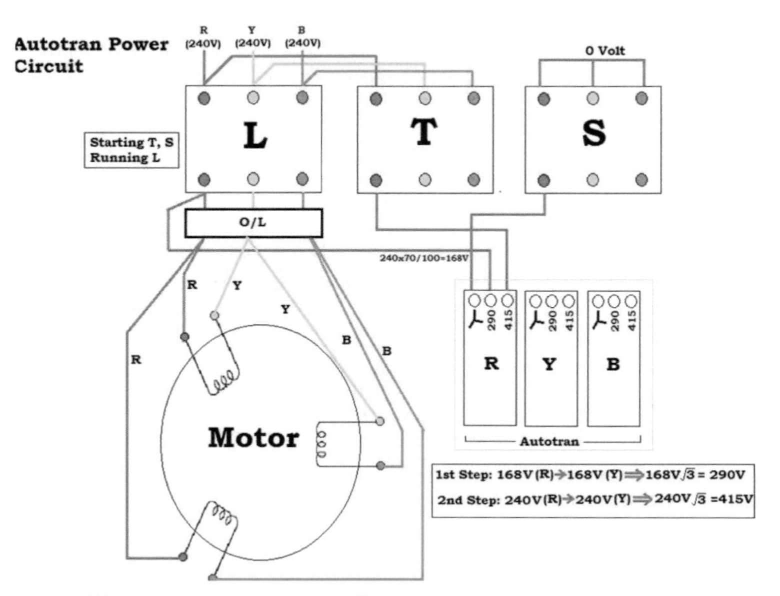


Fig. 44: Power circuit for Auto-trans motor starter

To achieve the above power circuit, the control circuit below is utilized:

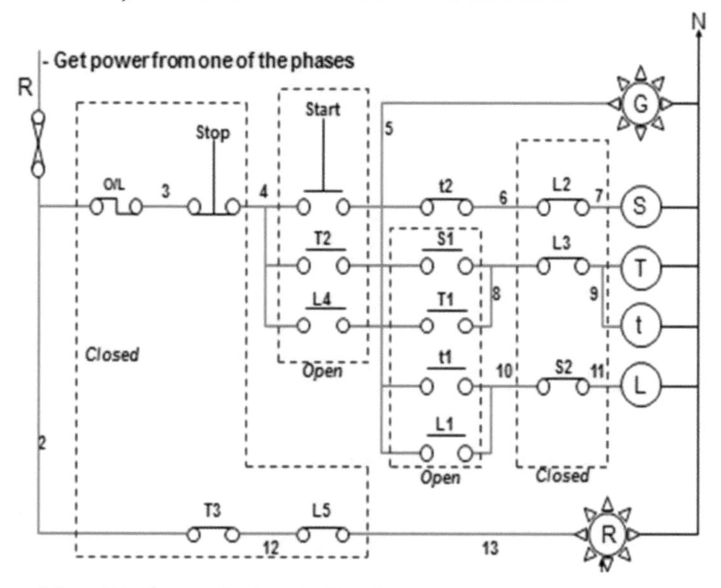


Fig. 45: Control circuit for Auto-trans motor starter

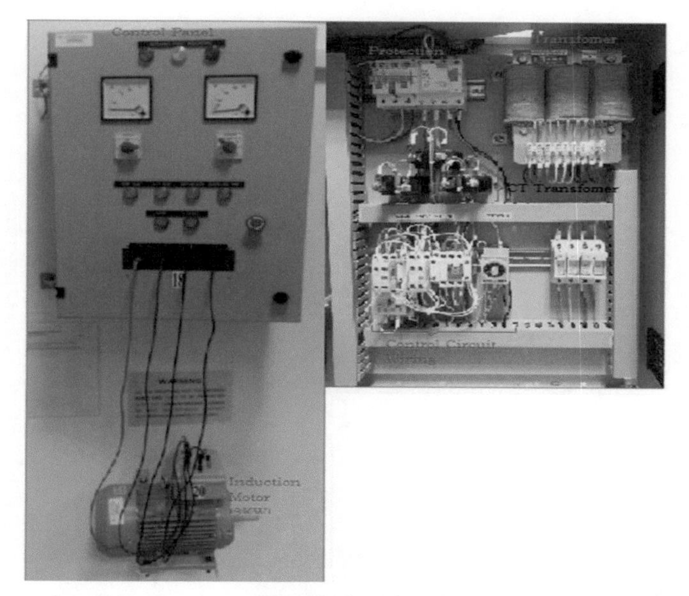


Fig. 46: The Control and Power combined circuit of an Auto- Trans-Delta motor starter

## 9.7 Variable Frequency Drive (VFD)

Currently many motors are being run with VFD but can cost 4000% higher and need an air condition unit to cool it down especially in hot and dusty factory environment. VFDs are also not as survivable and when if it fails another 40 times the cost (or 4000% extra cost) needs to be spent because it cannot be repaired. By building Star-Delta and Autotrans circuits one can realize a cost-effective method of starting motors. The circuit is simple and the parts are easily available and therefore serviceable. But VFD is only necessary if precise control of the motor is critical to the factory process.

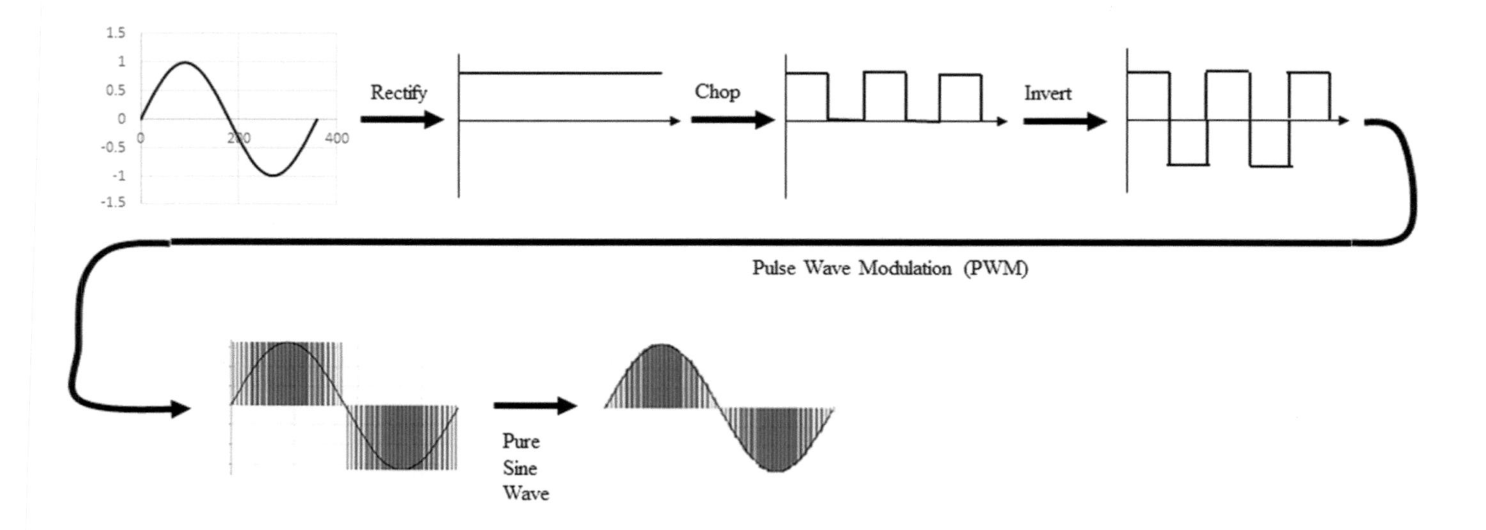


Fig. 47: Variable Frequency Drive (VFD) schematic showing the latest pure sine wave output

Variable Frequency Drives (VFD) or Variable Speed Drives (VSD) or Inverter is used to enable induction motors to have variable speed. It utilizes the Insulated Gate Bipolar Transistors (IGBTs) which was invented in 1982 as high-speed switches to achieve this. This is shown in Fig. 48.

# IGBT

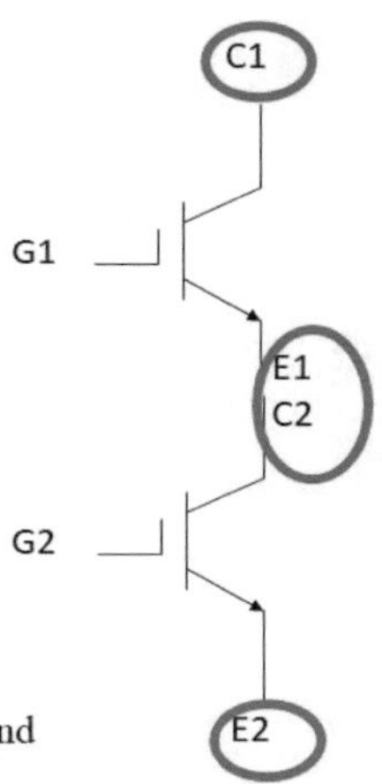


From left, the 1st terminal is E1C2 (big blue circle in schematic), the 2nd terminal is E2 (blue circle at bottom) and 3rd terminal is blue circle at top

Fig. 48: Insulated Gate Bipolar Transistor (IGBT) used inside a VFD to control a 25 HP motor.

It is actually two IGBT connected in series. The red connector is the Gate (G), takes in low-current signals from a microprocessor that switches On and OFF the 1st big terminal on the left. This 1st terminal cuts and opens continuity from the 2nd (incoming power) to the 3rd terminal (goes to the motor)

IGBT combines the gate drive characteristics of the MOSFET with the high current and low saturation capability of bipolar transistors (BJT). Thus, it has an isolated gate FET for the control input and a bipolar transistor as a switch in a single device.

In a VFD, line AC voltage at 60 Hz (50 Hz British) is first converted to DC and is chopped to give DC square waves. Six IGBTs are arranged in such a manner that these DC square waves are converted back to AC square waves with a variable frequency. The VFD is sometimes just called motor controller which is different from a motor starter like DOL, Star Delta and Autotran. Motor starters do not have controlling ability, it can only start the motor. A VFD enables running the induction motor at precisely controllable speed or torque. The most common VFD chops (switch on and switch off) the pulses of the same height (or same voltage) at a very high frequency. Then a square-wave of a particular period is formed. Over time labs around the world increase the frequency of chopping. These high frequency square waves are used to create waveforms for the motor to run. An analogy is if we use a two feet wide brick our home shape cannot change too much. But if we use a one-inch wide brick, we could make a very intricate shape wall or a "designer shape". That is the same logic why the frequency in increased to more and more perfectly form a sinusoidal waveform as shown in Fig. 47. Till today most of the VFD controlled motors are energized with square waves as shown on the second last graph in Fig. 47. So instead of the amplitude of the sine wave going up at the center, there are many more pulses of tiny square waves at the center and less on the sides. This is called Pulse Wave Modulation (PWM). The amplitude is fixed but the concentration of on time in increased at the center.

PWM is used in lots of industries. FM radio works on the same principle. Pure sine wave VFD uses inductors to output a varying voltage pulse at different times, thereby the output is back to sine wave which can increase the efficiency of induction motors by 15%. Note that in high power applications, inductors with taps in-between (Autotrans) are a preferred way to split up voltages just as a voltage divider resistor circuit does in low voltage. Therefore, a variable Autotrans in HV replaces the rheostat of LV.

The VFD is also called a VSD (variable speed Drive) and an inverter. This inverter term is wrong but is the most common term used worldwide mainly because that is the term used in ever increasing home appliances that uses VFD; including inverter air-conditioning, inverter washing machine and inverter refrigerators. Invert is electrical engineering term which means converting DC to AC which is the inverse of rectify which means converting AC to DC. In the inverter schematic shown in Fig. 47 above, converting from DC to AC is only one small portion of the process. Other processes are rectifying and computer control of the IGBT in performing the chopping. Thus, a VFD is a rectifier, a computer, a chopper and an inverter combined but is called an 'inverter'. Someone decided to use that term long back and it got stuck, just as the wrong term, 'Xerox' is generally used to denote photocopy in the USA. Fig. 47 is an upgrade of the normal inverter called pure sine wave inverter because the final output is a sine wave which is more suitable than the square-wave waveform which most VFD controlled motors in the world currently use.

Soon when the price of VFD goes down, it may even appear in home fans where there will be no more 1, 2, 3 speed; a tuner will be able to tune continuously to any speed desirable.

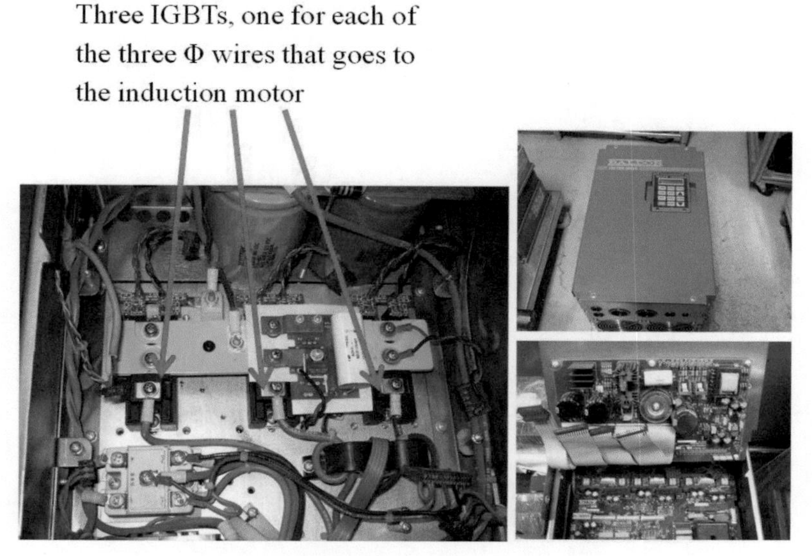


Fig 49: Outside and inside of a VFD

Upon opening up a VFD as in Fig. 49, one can see one or two large capacitors; these are used to produce the DC waveform. VFD causes one defect because chopping waves into square waves is not natural and causes lots of harmonics. Harmonics are the creation of many other frequencies of waveforms other than the fundamental power company frequency. This is especially so at the point of chopping a DC, that is, switching it on or off. This is at the cliff of the wave. At this point the amplitude of the harmonic can be so high that holes can be poked in the varnish insulation of the motor stator coils. Therefore, varnish of VFD controlled motors must be of a higher quality. So, do not think recoiling of an inverter-controlled motor can be done by normal recoiling shops which tend not to have such high-quality varnish around their coiling wires.

In building installations like hospitals (where tiny human body signals are measured) and telecommunication centers (where the bits are getting ever finer as time goes by), lots of problems can be caused by these harmonics. So, especially in these installations harmonic filters must be installed. A common harmonic filter is an isolation transformer which is a 1:1 isolation transformer. Isolation transformers can cut off 65% of harmonics. The best harmonic filters, called active harmonic filter can cut off 95% of harmonics. Table 3 is the list of available harmonic filters.

Table 3: Various types of harmonic filters in the market

| | |
|---|---|
| Active Harmonic filters | 95% |
| 18 pulse Drive System | 95% |
| Boradband Blocking Filter- Drives | 92% |
| Neutral Blocking Filter | 90% |
| Neutral Cancellation Transformer | 90% |
| Passive Harmonic Filter | 85% |
| Harmonic Mitigating Transformers | 85% |
| AC line reactors | 65% |
| DC reactors for drives | 65% |
| Isolation transformers | 65% |
| Neutral oversizing | |

In hospitals and telecommunication centers, Grounding (Earthing) must also be as good as possible, that is by using the best ground rods plus cad-welding or oxygen-acetylene welding. The cad-weld as shown in Fig. 50 will ensure a good bonding between the Ground (Earth) cable and the Ground (Earth) rod. Fig. 51 is the black power used as an explosive to achieving the melting of the copper. The black powder itself has a bit of copper in it. In the opinion of this author all Ground (Earth) rods should be cad-welded or oxygen-acetylene welded to the Ground (Earth) wire because the corrosion in-between the Ground (Earth) wire and the Ground (Earth) rod can render the whole system of Grounding (Earthing) useless. Corrosion is copper oxide or iron oxide. Iron oxide is at the joint because most common Ground (Earth) rods are made of iron and only plated with copper. Considering the price of the long Ground (Earth) green wire and the expensive high quality Ground (Earth) rods and the manpower to install the whole system, it is worth it to invest in a cad-weld or oxygen-acetylene welding equipment to make a proper join. That is, the bonding of the green ground wire to the ground rods is the weakest point in the whole system. Corrosion is the formation of metal oxides in this case copper or iron oxides. All metal oxides are insulators. In the experience of this author most contractors serious about having a good join between the copper grounding wire and the grounding rod use oxygen-acetylene welding because the cad-weld equipment needs to be held in place on top of the ground rod which nobody does because everyone knows the black powder is gun powder so run away as it is ignited, so, the cad-weld equipment falls a little or totally during the explosion. But using an oxygen-acetylene flamer and a copper welding rod (probably mixed with tin) which is commonly found in hardware stores, the human hand can precisely control the welding at the right spot.

Fig. 50: Cad-welding of copper strip which is joined to Ground (Earth) rods for a telecommunication tower

Fig. 51: Black power used to do cad-welding

## 9.8 Synchronous motor

The basic structure and operation of a synchronous motor is very similar to an induction motor. The main difference in a synchronous motor is that the rotor is not just an un-energized squirrel cage but has coils through which DC is sent to via carbon brush. The workings of a synchronous motor are similar to an induction motor except that the rotor is turning synchronously with the rotating magnetic field (RMF) of the stator. In an induction motor there is a difference in speed of rotor versus RMF called slip. But at the starting point, the rotor has inertia and cannot attach to the RMF with 'strings' of magnetic field fast enough. A N pole is initially attracted to a S pole of the RMF will later slip to be in front of a S pole. Thus, a rotor coil will experience attractive force to the RMF and then repulsive force. Therefore, the motor cannot turn or synchronous motors are not inherently self-starting. To overcome this problem, initially the DC to energize the rotor coils are not switched on. A squirrel cage is placed above the rotor coils. This squirrel cage will enable the synchronous motor to be self-starting just like an induction motor. This squirrel cage will carry the coiled rotor along. Once the rotor has reached almost synchronous speed (minus a slip as in an induction motor), the DC to the rotor coils will be turned on. Now the rotor becomes a solenoid and a N pole of the rotor coil will be latched onto a S pole of the stator coil and the synchronous motor will thenceforth rotate synchronously with the RMF.

The DC goes into the rotor via a pair of slip-rings and carbon brushes, making the rotor behave like a concentric array of bar magnets. Another option is to use a permanent magnet as the rotor. The need for slip-rings and carbon brush are the main disadvantage of the synchronous motor.

All generators are synchronous motors. A recent use of synchronous motors is to replace stepper motors in robot arms. With stepper motors, we could mechanically switch the energizing of coils in simple steps as shown in the stepper motor of Fig. 52. In step one, the top coil is energized to be a N then the side coil is energized to be N then the top coil is again energized with a S and then the side coil is energized as a S. With this capability, robot arms were actuated almost solely with stepper motors. Robot engineers like this because the operation of the stepper motors is like a computer, step-by-step of digital in nature. But today a software called DSP (Digital Signal Processing) which was developed by the telecommunications industry is being used to run synchronous motors with lots of stator coils. Such a synchronous motor can be utilized to move robot arms. The more coils in the rotor, the more precise will be the robot arm's movement capability. DSP software can instruct the energizing of coils precisely so synchronous motors can be used. The advantage of synchronous motors is that they are much more powerful than stepper motors. Most stepper motor actuated robot arms can easily be stopped by a human hand whereas there is actually no limit on the power of a synchronous motor. The current largest motor in the world is the 135,000 HP synchronous motor build by ABB for a NASA wind tunnel.

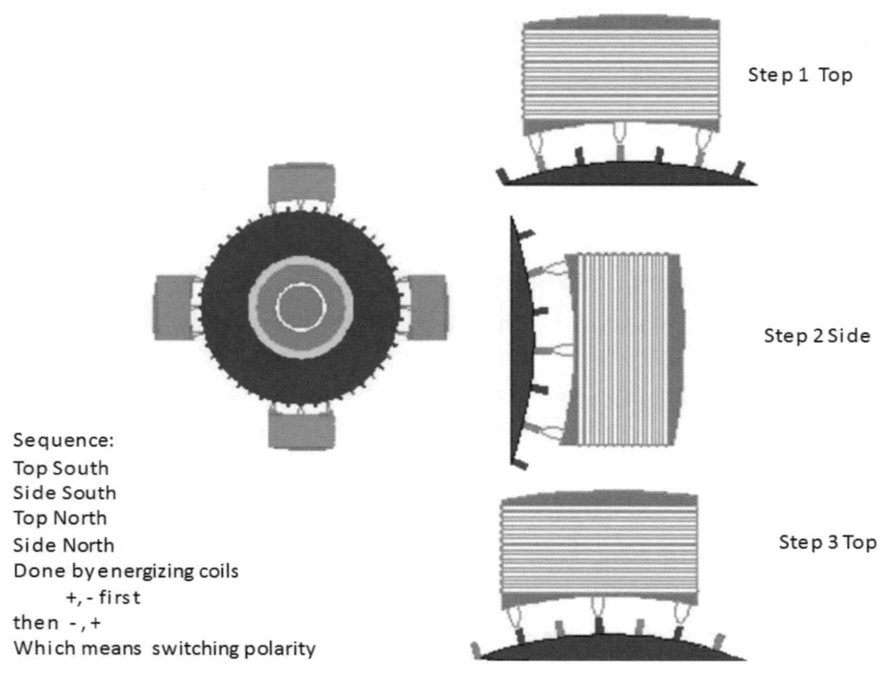


Fig. 52: Stepper motor functioning sequence

## 9.9 Brushless generators

As mentioned previously all generators are synchronous motors. A brushless generator is actually two generators on one shaft. Smaller brushless generator's shaft looks like one unit but the two parts are more easily seen on larger brushless generators. The larger of the two sections is the main generator and the smaller one supplies DC current to the rotor of the larger generator. The small generator therefore acts as the exciter mechanism for the larger generator. The smaller generator has a stationary stator coils through which DC current is sent. This will induce current in the rotor coils below it. The consequent AC generated in the rotor coil below is sent to the shaft in-between the small and big generator. Here the AC is rectified by a rectifier mounted on the shaft. The body of a brushless generator of sizes 20-40 HP has a hole at the point of the shaft where the rectifier is mounted onto the shaft. This way, convection air can cool it down. The resulting DC is sent to the coils of the bigger generator's rotor. Here the coils become electromagnets whose field lines cuts the coils of the stator above generating AC current. The biggest problem with electronics (rectifier in this case) is the need to cool it down to operate efficiently. Only recently have electronics been invented which can withstands the temperature and dust of a motor shaft, which is why this simple idea is quite new.

## 9.10 Universal Motor

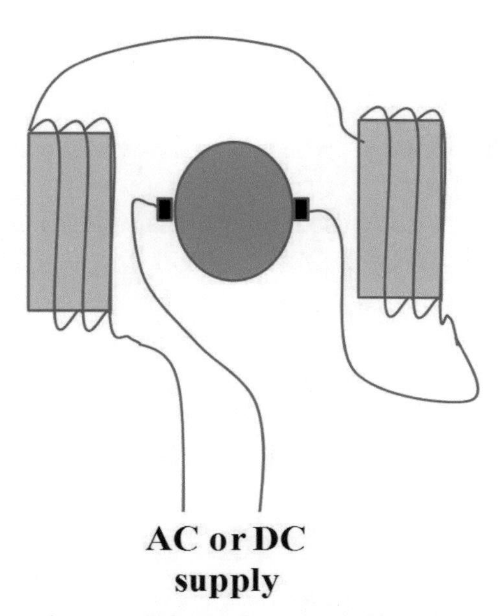


Fig. 53: Universal motor schematic

This motor is named universal because it can run on either AC or DC power sources as shown in Fig. 53. Fig. 54 is how the drill and most power tool's universal motor looks like. They have a high horsepower-to-size ratio. Among the machines using this motor are vacuum cleaners, food mixers, portable drills, portable power saws, and sewing machines. These motors are rated one HP and below and do not turn at constant speeds, the speed varies with the load. The higher the load the slower it turns which is exactly what is needed when cutting a piece of iron for example. A soft iron piece can be sawed fast with a hacksaw but a hard piece of iron needs to be sawed slowly. So, when drilling soft concrete, the drill speed is fast and when hard concrete is encountered, the drill bit will automatically slow down. When the motor operates with no load, the speed may attain 15,000 RPM. With a heavy load the motor runs at a few hundred RPM. The rotor of the universal motor is made with laminated iron wound with wire coils just like an induction or synchronous motor. Electric current flows in the stator windings as well as the rotor windings via a carbon brush. The direction of coiling is opposite in the rotor compared to the stator. Say looking at the stator coil from the point of the rotor it is anticlockwise. To get the rotor to have the same magnetic pole at that moment the coiling has to be in the opposite direction (looking at the rotor from the stator). If DC is sent to the stator, it will travel to the rotor coils also. The rotor will have many more solenoids than the stator. At one moment, one rotor coil will be a

N, the next a S and the next a N and so forth. The N will be attracted to a S stator pole and the next rotor pole is a S which will repel the stator pole but the first action above has already pushed the rotor enough that this repulsive force enhance the rotational movement. Using AC, a positive pulse travels through the stator coils as well as the rotor coil at the same time providing a case as above for the DC case. When the AC switches direction, the rotational movement continues via the attraction and repelling forces.

For electrical contractors it is better to purchase a higher quality drill alike De Walt or Hitachi. There is a very significant difference between a 300-watt drill and a 750 watt one. The difference can be a few seconds per hole versus up to half hour per hole. With the few seconds per hole drill, man-hours and therefore output per worker can be drastically improved. For bad quality drills, the outside looks the same but this author has seen the electric motor shaft being held by the plastic casing; without a bearing. For these bad drills, the whole housing of the motor is made of plastic while a Hitachi has the motor housed in a metal housing with good bearings at both ends. Fig. 54 is a Hitachi drill whose carbon brush lasted from 1993-2018 = 25 years.

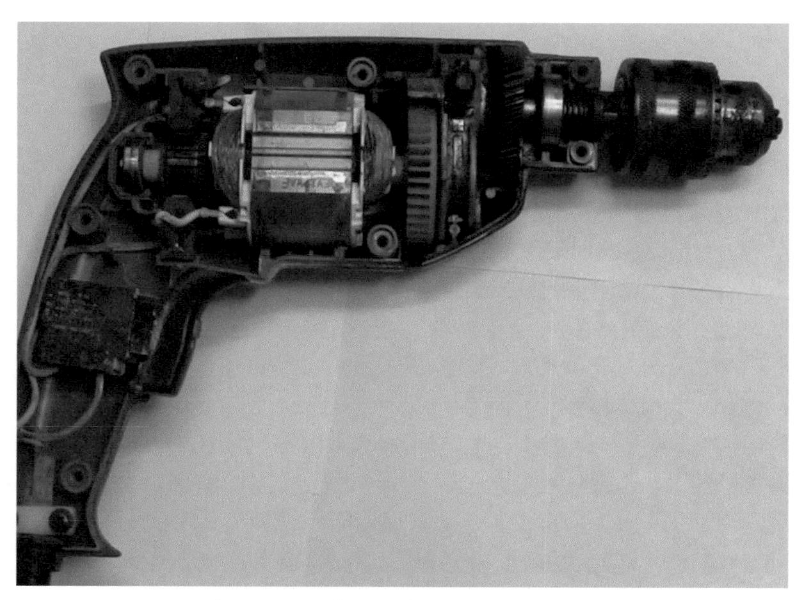

Fig. 54: Universal motor in a drill

# Chapter 10

# Inductors

Inductors are simply coils of wire which behaves as a magnet as shown in Fig. 55. Inductance is the property of generating an electromotive force (voltage) to oppose a change in current flow. The following story helps in understanding inductors. The nine-year-old son of this author asked for a piece of copper wire, a nail and a power supply so he could make an electromagnet as described in a science experiment book he had just read. This author knew inductors coils always have varnish insulation around the copper wire but the young son said it will not be necessary. What was given to him was one strand of copper wire from a 1.5 $mm^2$ cable, a 12 V car battery charger power supply and a three-inch nail. The son just coiled the bare Cu wire around the nail and connected the two end terminals to the battery charger. The nail turned into a powerful magnet. But the bare wire did not blow up the fuse in the power supply; though it was much hotter than a solenoid with varnish insulated Cu wire.

Comparatively if the end of a length of uncoiled bare copper wire is connected to the terminals of the battery charger, it is a direct short-circuit and the fuse and other component in the battery charger will blow up. As mentioned previously a short-circuit can go through protective devices and damage components because it can rise to kilo amperes. But in this case the same bare Cu wire is shaped into a coil and there is no short-circuit. Why is happening here? When a bare wire is shaped into a coil, it becomes a solenoid and therefore a magnet. It can be imagined that a magnet is placed above a wire, which restricted the flow of electrons in the wire. This is therefore a load and there is no short-circuit.

Joseph Henry who invented the first practical solenoid went through the same process, just as in this author's son's experiment, the solenoid without insulation around the coils will get very hot. So, it was not practical to make more powerful solenoids. So, he insulated the coil of copper wires by coiling silk string around it. This must have taken a long time for him to do. Today, varnish is the standard. The range of varnish is large and the high quality, stable and long-lasting motor, generator or transformer is primarily determined by the type of varnish around the copper coils. The other factor for quality is the purity of the copper used in the winding.

It must be noted that some electricians in industry like to coil electrical wires before a termination point; they do this because in future they have this spare length of wire to cut and redo the termination. They can do this but the wire cannot be shaped as a coil. Just remember coils are sensitive to electrical flow. It can be shaped into a long S if needed.

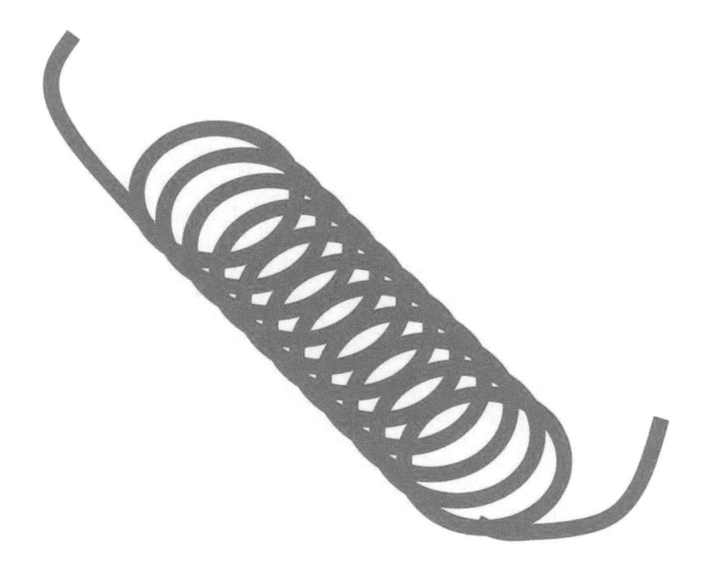

Fig. 55: Inductors

Table 4: Charging and discharging of an inductor

| Time | Battery voltage | Inductor voltage | Current |
|---|---|---|---|
| 0 s | 15 V | 15 V | 0 A |
| 0.5 s | 15 V | 9.098 V | 5.902 A |
| 1 s | 15 V | 5.518 V | 9.428 A |
| 2 s | 15 V | 2.030 V | 12.970 A |
| 3 s | 15 V | 0.747 V | 14.250 A |
| 4 s | 15 V | 0.275 V | 14.730 A |
| 5 s | 15 V | 0.101 V | 14.900 A |
| 6 s | 15 V | 37.181mv | 14.990A |
| 10 s | 15 V | 0.681mV | 14.990A |

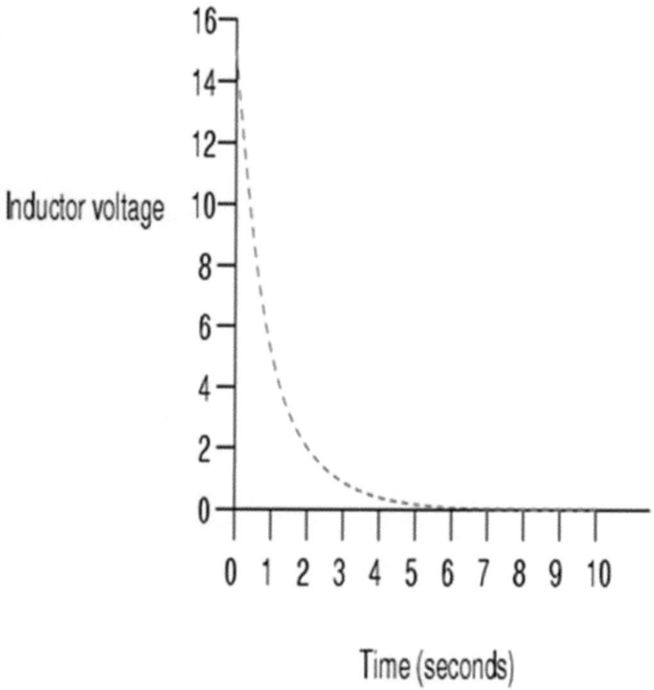

The end of the coil where the electrons flow (opposite of current flow) is in a clockwise direction is the N and the other end is the S of the magnet (somehow the human mind can more easily remember and relate clockwise to N so electron flow needs to be used). The coil needs to have a layer of insulator normally a thin layer of high-quality varnish. The original varnish was from a tiny parasitic lac insect (Laccifer lacca) found on certain trees in India and Thailand. The electrical varnish used today is totally created in labs and manufactured from petroleum and other hydrocarbons. One to four layers of varnish insulation is used. Different films are used for different heat resistance, namely (in order of heat resistance) polyvinyl formal, polyurethane, polyamide, polyester, polyester-polyimide. Electrical engineers give the specification to chemical engineers to develop the varnish and the specification are, excellent flexibility (in transformers the laminates in the core need to vibrate), heat shock resistance (transformers and motor coils must withstand sudden high heat), thermal life (transformers and motors need to operate up to 60 years), and abrasion resistance (the laminates in transformers and motors may slide over each other). Most varnish is made to evaporate with the heat of soldering iron to ease joining.

As shown in Table 4 inductors store charge for a while before discharging it like a capacitor. In this case it almost completely discharges its magnetic field (which provides voltage) in about eight seconds. After discharging it is more like a short-circuit with a voltage of 0.681V; note voltage across a component is proportional to the resistance across it, in this case inductive reactance. So, at 10ms, the coil of wire becomes has almost no magnetic field around it (no resistance to electron flow), so the current shoots up to 14.99 amps.

Inductor's (solenoid) magnetic strength increases with current. A physical experiment with a FLUKE 1AC-E II VoltAlert detector indicates this. It only lights up and makes sound when it is placed on the L wire but not on the N wire. If it is placed on the circumference of a three-core single-phase flexible wire like a computer incoming wire, it will light up only where the L wire is. The L wire portion of a three-core wire is the only portion of the circumference which has voltage. The difference between the L wire and the N wire is only the voltage; the current must be the same but in opposite direction or else the incoming RCD will trip. This equipment detects electrostatic forces and not magnetic field. It basically detects ions which are which are influenced by the presence of a voltage. Magnetic fields increase proportionally with the current flowing in the wire whereas electrostatic forces are the attraction between electrons and protons or attraction between positively charged ions and negatively charged ion. Light or electromagnetic waves are not affected by magnetic fields but light is refracted by atoms that are moving at different speeds; just as light is refracted by water.

The basic formula for field strength is:

$$B = k_1 NI \qquad (25)$$

AC current generates eddy current which pushes all current in an AC wire to go the outer diameter portion of the cable. The formula for eddy current is as below.

$$P = \frac{\pi^2 B_p^2 d^2 f^2}{6kpD} \qquad (26)$$

Equation 25 shows that magnetic field strength is only affected by I and equation 26 shows that the eddy current loss is affected by B which is connect to I in equation 25.

When designing a solenoid, start with the current carrying capacity of the wire. Then divide the supply voltage with this current to get the required impedance. From the impedance required (say Z Ω) the number of turns required can be calculated to achieve the required impedance. Say ten turns is first made and the impedance measured with a multimeter (say Y Ω) then the

$$Y\ \Omega \text{ --- } 10\ N \text{ where } N = \text{one turn} \quad (30)$$

$$1\Omega\text{---}\ \frac{10N}{y} \quad (31)$$

Therefore, $Z\ \Omega$ --- $\frac{10N\ X\ Z}{y}$ turns (32)

Which are the required turns needed to achieve the impedance required.

As mentioned previously the slowing down of current flow imposed by an inductor is called inductive reactance and follows the equation 33. Electrical technicians and engineers need to memorize this because it is useful in industry and it also comes out in electrical certification examinations. Since this formula is similar to the capacitive reactance formula, the way to remember it is that this formula is in one line; the L in the formula means it is in one line and the capacitive reactance formula is in two lines.

$$X_L = 2\pi fL \quad (33)$$

# Chapter 11

# Capacitors

Capacitors are basically two plates with a dielectric in-between them. The formula for a capacitor is:

$$C = \frac{\varepsilon_o \varepsilon_r A}{d} \quad (34)$$

Where: $\varepsilon_r$ = Permittivity of the material

$\varepsilon_o$ = Permittivity in free space
A = Area
d = distance between plates

To increase the capacitance value, the A in-between the plates can be increased or the value of d between the plates can be reduced. In a high capacitance capacitor, the A is increased by coiling two sheets of aluminum foils with a dielectric in-between which will enable achieving a higher A. The best dielectric is mica which is found naturally in the ground in certain spots of the earth. With mica as the dielectric the separation between capacitor plates can be as short as possible, thereby an even longer coil of dielectric sandwiched between aluminum sheets can be made to achieve a very high capacitance. Mica is preferred as a dielectric because of its high dielectric strength and excellent chemical stability. The dielectric, mica enables the A to be large and also the d to be small to achieve high capacitance. Fig. 56 and Fig. 57 shows the outside and inside of a capacitor. Recently the word dielectric strength has become a replacement for insulation strength in most electrical books and journals. This is just like chemical engineers using the word polymers to replace the common word plastic.

Fig. 56: Capacitor

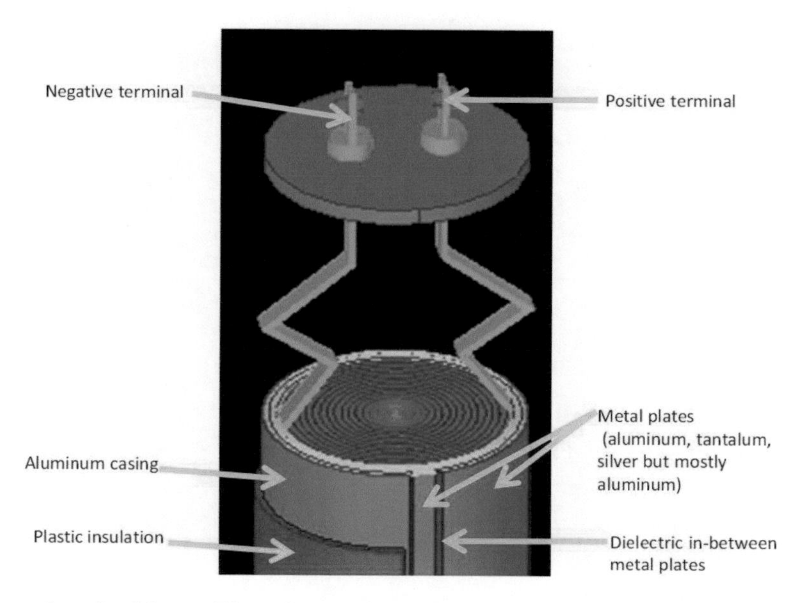


Fig. 57: Capacitor inside, coiling the two metals with dielectric in-between increases A

For a capacitor the formula is:

$$I = C\frac{dV}{dt} \quad (35)$$

When dV = 0V no current flows

For an inductor the formula is:

$$V = I\frac{dI}{dt} \quad (36)$$

When dI = 0 no voltage is experienced

# Chapter 12

# Transformers

Michael Faraday deduced the Faraday's law of induction which is now the fundamental principle of electromagnetism and transformer operation. He invented the transformer in 1831 after inventing the motor and the generator soon after. Therefore, he invented the most fundamental devices required for the electrical world we live in now. Faraday wrapped two wires around opposite sides of an iron doughnut and found that if he passed an AC current through one, it immediately induced a current in the other.

Functions of a transformer are:

1) Step-up step-down of voltage and current
2) Isolation – for protection of controllers
3) Impedance matching
4) Remove DC components from AC waves
5) To transfer AC signal without using brushes or other moving contacts
6) Get rid of noise and spikes

However, transformers are mostly utilized to raise or lower voltage. It has no moving parts and is simple, rugged, and durable. A transformer is the most efficient machine in the world; low power transformers can have efficiencies of 80-90% while high power line transformer can have efficiencies of up to 98%. This means the energy going out of a transformer over the energy going in is 98%. Comparatively the energy of a car movement over the energy inherent in the petrol is typically only 25% (maximum of 37%).

A transformer is basically two inductors placed next to each other with an iron core connecting them. It uses the property of electromagnetic induction to get current from one coil to induce current in the second coil other without any electrical wire connection. The coil with the incoming electricity is called the Primacy coil and electricity comes out of the Secondary coil. If the transformer is turned around the Secondary will become the Primary and so forth. But it is critical to ensure the current carrying capacity of the coils in the transformer can handle the load. The wire size of the coils is determined just like any other wiring where the current drawn must be calculated and the appropriate wire size is chosen from the Cable Table.

In 1905 paper, Albert Einstein established relativity which showed that both the electric and magnetic fields were part of the same phenomena viewed from different reference frames. For the magnetic field lines to travel from the Primary coil to the Secondary coil, the best medium is iron. It needs to be repeated here that the conduction of magnetic field is best in iron. So, iron has the best permeability. Permeability is always taken as the permeability of the element of material over the permeability of air. Permeability of air is 1. It turns out that the permeability of most other elements is also 1. This includes copper, silver, aluminum and gold, wood etc. In other words, iron is the best conductor of magnetic field lines, just as silver is the best conductor of electrons (with copper being number two and aluminum number three).

As an indication of utility of the iron core, this author has tried taking out the iron core from the solenoid of a contactor and energizing it; there is magnetic pull for a steel screwdriver placed near the coil but it is very weak. Placing the iron core back in the center of the coil turns it to a powerful magnet for the screwdriver placed nearby. What is happening is that magnetic field lines can travel easily through the iron thereby greatly increasing its magnetic strength.

Inductance is the property of generating an electromotive force (voltage) to oppose a change in current flow, the change being the AC which causes a variation in magnetic field density. Though iron conducts electricity, in transformers it is only used to conduct the magnetic field lines from the Primary to the Secondary. Current flowing in the varnish insulated wires of the Primary will generate magnetic field lines around the Primary coil. This magnetic field lines are continuously changing direction because AC current is sent to the Primary. A varying AC current creates a varying number of field lines which is a prerequisite for current induction in the Secondary coil. If DC is sent to the Primary coil, it will be a strong magnet with lots of field lines but it will not be varying so there will be no current induction in the Secondary coil. But when the DC is first switched on there is a change in number of field lines from zero to full so a bulb joined to the Secondary will light up once and after that it will never light up even if a powerful DC is flowing in the Primary (resulting in a very high magnetic field density). As already explained earlier in this book, what is happening in a transformer is exactly as what is happening when a magnet is passed a copper wire. When the magnet is initially far away from the wire, it will “see” only a few field lines, as the magnet get closer to the wire, it will “see” and increasing density of field lines and as it crosses the wire, it will “see” a maximum field line density and the density reduces as it moves away from the wire. So, what the wire is “seeing” is a change in magnetic field density. In the what is taught in schools, magnetic induction follows F, B, I (remembered as the American secret service which has guns so placing the fingers like a gun, from top to bottom is FBI). In schools F is taught to be Force, B is magnetic field (N at the beginning of this finger and S at the end) and I is the direction of the current. But this author prefers to call F varying-**F**ield-density. This way induction of current in a wire of generators, motors, transformers, skin effect in AC wires can all be explained at once. Transformers can only work with AC because the voltage and therefore the current (which lags a little) is varying. Note as previously explained in this book, it is the current that creates magnetic fields. In AC the current is initially zero, then goes up the sine wave to the peak, then goes down the sine wave and reaches the negative peak and eventually reaches zero again. The magnetic field density varies in proportion to the current and therefore the magnetic field density follows the equivalent sine wave. So, there is a varying field density which enables current to be induced in the Secondary coil. Even if a huge DC current is sent to the Primary coil, there will be no induced current in the Secondary because there is no variation in the magnetic field density.

On why current only flows in the outer portion of an AC cable (skin effect). Electrons are equal to tiny magnets. As these tiny magnets start moving in one direction, it achieves maximum speed at the center and reduces speed till it stops at the end of this direction. It then goes back other direction to where it started initially. As the speed of electron flow increase and peaks the center of this back-and-forth direction, the magnetic field density variation also increase to a peak as the electron velocity is highest and then reduces to reach zero at the other end. Therefore, there is magnetic field density variation, which means current generation will happen. Thereby there is current generation in an AC carrying wire and following through with your fingers and the right-hand rule, will show that this current generation is outward of the center of the wire. This is an unwanted current generation and any unwanted flow of current (for human machinery to work) is termed eddy current. Imagine each electron to be a magnet where the N and S are oriented in multiple directions. Using Fleming’s Right-Hand rule, this will cause current generation in a multiple direction but all out of the center of the wire. The overall effect is to cause a pipe like flow of electrons near the surface of the wire. The eddy current generated also pushes all the electrons generated by the big generator far away to the outer diameter of the wire. This moving of electrons only at the outer skin of the wire when AC is sent through it is termed skin effect. In overhead grid cables, only about 8mm thickness from the outer diameter is used for conductance of electron. So, at the center of overhead lines, steel strands are placed for mechanical strength. It should be noted that the frequency of all grid frequency is 50 Hz (Britain) or 60 Hz (USA) but in airplanes it is 400 Hz.

The reason why they increased the speed of rotation in generators of airplanes is because at higher rotor speed there is higher variation in magnetic field density resulting in higher current generation. Thus, higher current generation can be achieved with a smaller size and lighter generator which are critical factors in airplanes.

Thereby the magnetic field density is also proportionately varying with the speed of the electron. There are three losses in transformers, namely eddy current losses, hysteresis losses and copper losses. Transformers are rated in VA or kVA. It is just fashion or convention. But in a Pythagoras theorem, there are four things the angle, the line opposite the angle (opposite) and the line adjacent to the angle (adjacent) and the longer third line called the hypotenuses. Among these four things as long as two is known, the other two can be calculated.

Transformers including CT (Current Transformers) or Voltage Transformers (VT) or Potential Transformers (PT) are all rated in VA units. Motors are generally rated in Watts or HP and capacitors are rated in Vars.

A transformer is mainly used for stepping up or stepping down voltage or current, following the formula:

1,2, 1,2, 2,1

$$\frac{V_1}{V_2} = \frac{N_1}{N_2} = \frac{I_2}{I_1} = \sqrt{\frac{Z_1}{Z_2}} \quad (37)$$

$$\frac{V_p}{V_s} = \frac{N_p}{N_s} = \frac{I_s}{I_p} = \sqrt{\frac{Z_p}{Z_s}} \quad (38)$$

Where $N_p$ is the number of turns of the primary coil and $N_s$ is the number of turns in the secondary coil. $V_p$ and $V_s$ are the primary and secondary voltages, $I_p$ and $I_s$ are the primary current and secondary currents and $Z_p$ and $Z_s$ are the primary and secondary impedances.

In this author's experience of teaching this subject, memory of this formula is an Achilles heel for many students, who tend to forget it during examinations. A slight mistake in memory can cause hugely wrong results. This brings to mind that electrical engineering is not like a math class where we can get 50% and pass or even 95%. In electrical engineering we need 100% or the effect can be an exploded transformer or so forth. So, a three-step method was devised to remember the formula. The first step is remembering 1,2,1,2,2,1; like remembering a phone number, then $V_1$ over $V_2$ equal to $N_1$ over $N_2$ equals to $I_2$ over $I_1$ and finally $V_p$ over $V_s$ equal to $N_p$ over $N_s$ equals to $I_s$ over $I_p$. Perhaps because the words Primary and Secondary are long words, they do not stay in memory easily. It is easier for humans to remember numbers like a phone number.

As mentioned previously transformers are used to step up voltages from the 11-13kV output of generators to 275 kV or 500kV used for transmission. This will reduce the current. Typical current in 275 kV lines range from 10 to 59A; as a comparison, a typical car battery outputs 70A upon startup. But it is safe to touch a car battery terminal even during startup because the voltage of the battery is only 12V which is below 40V required for current to jump into a human skin. Note if current does enter the human body, 1A is enough to kill; the 70A at startup of a car does not kill because voltage is 12V.

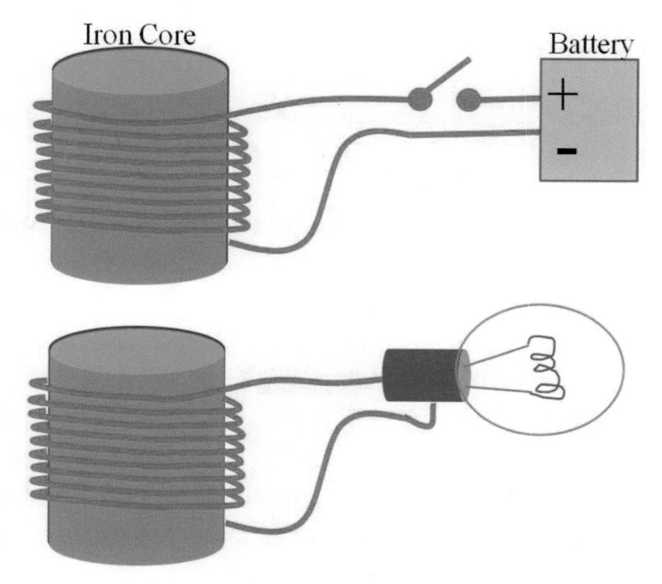


Fig. 58: Transformer logic

Looking at Fig. 58, if the switch is switched ON, the bulb will light up once and will not light up again. What is happening is that the current in the Primary increased from zero to high resulting in a proportional change in magnetic Field density from zero to high. This change in field density generated current in the secondary causing the bulb to light up once. After that, it will not light up again. So, to get the bulb to light up again, switch on and off the switch again. Therefore, to get the bulb to continuously light up, switch on and off the switch very fast. Alternatively, the battery can be replaced with AC wire of a home. What is happening in

AC is, say the L wire of the home is eventually connected to the L1(R) phase coil in the generator 1000 km away. At the power company generator, the rotor magnet just before crossing the L1(R) phase coil is switched off, as the rotor magnet passes the L1(R) coil, the home AC wire will be switched on and then off again as it passes this coil. So, in the 1000km away generator the off and on is being done so the bulb continuously light up. In effect, the far away generator is an on-off switch replacing the fast-moving hand to switch on and off the DC supply of the Primary coil.

The fact that induction of current requires a varying Field density is analogous to F = ma where a varying speed called acceleration (a) is necessary to achieve a force, a constant speed do not result in a force. In a transformer, a varying Field density (analogous to acceleration) is needed.

## 12.1 Losses in a transformer

As mentioned previously there are three losses in a transformer, eddy current loss, hysteresis loss and copper loss. A deeper description of these will be given in the next few paragraphs.

The function of the iron core in a transformer is to transmit magnetic field lines. But iron is also a conductor of electrons. The varying field density of the Primary not only generates induced current in the Secondary, it also induces current in the electron-conductive iron. But the iron core does not have any polarity like the Secondary of the transformer so the induced current flows like a hurricane of electricity within the core. Just as a hurricane is created by a variation in temperature and pressure in nearby places, a storms of electron flow is created within the core iron by voltage and heat. And any electron flow generates heat and this heat loss is termed eddy current loss. Eddy current in the core iron causes other problems for the transformer because eddy currents travelling like a hurricane will form their own magnetic field lines which will interact with the magnetic field lines of the coils thereby interrupting the transformation function. The resulting AC waves coming out of the transformer is not a clean sine wave and will have distortions caused by the eddy current. The solution to eddy current loss is to use laminates of silicon steel, with varnish in-between. This way the eddy currents can only circulate within one laminate; thereby reducing them drastically.

The next loss is hysteresis loss. Domains are found in all ferromagnetic materials. Iron is used as the base of most magnets, since it is the cheapest ferromagnetic material. Within ferromagnetic material like iron, tiny regions called domains exists where unpaired electrons spins align making each domain a powerful magnet. But because the unpaired electrons spin in each domain is oriented in different directions, the overall magnetism in iron is zero. But after swiping a magnet over a piece of iron needle, as the young son of this author did for forty times (following a science guide book), the iron needle became a magnet. What happened in the needle is that the unpaired electron spin direction of all the domains started flow in the same direction; turning the steel (95% iron) needle into a magnet. In a transformer, as the power company 60 Hz (50 Hz British) AC is sent to the Primary, the magnetic field lines changes coinciding with the voltage changes. That is, magnetic field flows say in one direction to a peak, goes to zero and then builds up in the opposite direction to a peak and goes to zero again. The transformer core is silicon steel (95% iron); so, as the sinusoidally varying current is sent to the Primary and induced in the Secondary there will be electron changing direction of spinning following the AC frequency. This changing in direction of spinning at two times the generation frequency generates heat. So, the change in direction of spinning is 120Hz in USA and 100Hz in Britain. This heat loss is termed hysteresis loss. The solution is to use silicon steel as the core. Silicon steel means some silicon or sand ($SiO_2$) is added to the steel to reduce the number of domains within the iron. The silicon must be added carefully into molten iron to form the silicon-steel used in the manufacture of transformers. If too much Si is added it will render the iron core incapable of conducting the magnetic field lines from the primary coil to the secondary coil (note iron is the best conductor of magnetic field lines). And adding too little will cause a buildup of heat in the core by hysteresis loss. One senior contractor in this author's class was asking, "If silicon steel of transformers is just adding sand ($SiO_2$), why is it sold for such a high price?" The high price is due to the accuracy at which sand must be added to the molten iron. If too much is added, the conductance of magnetic field (permeability) is reduced too much and if too little is added, hysteresis heat loss will increase.

Start with an initial voltage, say V = 240V, choose current say I = 50A

$$V = IR \quad (44)$$

$$\frac{V}{R} = R \quad (45)$$

$$\frac{240}{50} = R$$

$$4.8\Omega = R$$

For 50A a 16mm$^2$ will which can handle 68A will suffice. Try 10 turns of 16 mm$^2$ wire plus a 40W incandescent bulb (which measures 106 Ω) as a load. Now there is a coil in series with a 106 Ω load. Now the current in the wire can be measured with a clamp meter. The voltage across the coil can be measured with a voltmeter. With the values of the current and voltage the resistance across the coil can be calculated as:

$$R = \frac{V}{I} \quad (46)$$

Then by the following calculation the number of coils needed can be calculated as:

$$10T --- 0.2\Omega \quad (47)$$

$$\frac{10}{0.2} --- 1\Omega$$

$$\frac{10}{0.2} X 4.8 --- 4.8\Omega$$

$$= 240\, Turns$$

## 12.2 Autotrans

An autotran is a single coil transformer with a tap in-between. It has a limit of a 3:1 ratio transformation. So, it cannot be used as a RMU (Ring Main Unit) transformer where the voltage transformation is 11kV to 415V.

$$\frac{11kV}{415V} = \frac{27}{1}\ or\ a\ 27:1\ ratio \quad (48)$$

Autotrans are used mostly as a motor starter or to change one country's voltage to another. For example:

Malaysia (240V) to USA (120V) voltage is:

$$\frac{240V}{120V} = \frac{2}{1}\ or\ a\ 2:1\ ratio \quad (49)$$

$$or\ \frac{1}{2} = 50\%\ of\ Malaysian\ voltage$$

So, to run a USA equipment in Malaysia an Autotrans can be built and if it is a 100 turns Autotrans, at 50 turns the varnish is scrapped off and a tap is soldered to this point.

Malaysia (240V) to China (220V) voltage is:

$$\frac{240V}{220V} = \frac{1.09}{1} \text{ or a 1.09:1 ratio} \quad (50)$$

$$\text{or } \frac{1}{1.09} \approx 90\% \text{ or Malaysian voltage}$$

So, to run a China equipment in Malaysia an Autotrans can be built and if it is 100 turns, at 90 turns the varnish is scrapped off and a tap is soldered to this point. Of course, in both cases the wire size must be chosen in accordance with the amps drawn by the equipment. For example, for a machine that draws 50A, 16mm$^2$ wire can be used because this wire can handle 68A.

Autotrans and tapping of Autotrans are used widely in electrical power systems, most voltage drops to enable the use of advance power electronics like IGBT (Insulated Gate Bipolar Transistors) or thyristors uses Autotrans. For example, one of the latest IGBT can handle 3.3 kV but how to switch a 11kV housing incoming line? A series circuit of four Autotrans can be built and this will cause a voltage drop of 3.3 X 4 =13.2kV. So, a 3.3 KV IGBT switch is placed parallel after each coil. Then the output of all the IGBT is combined to get back the 11kV. So, basically four autotrans and four IGBTs are used to switch on or off the 11kV. This way each IGBT switch will have to switch only 2.75 kV which is less than its capacity of 3.3 kV. This way even 500kV can be switched with even more autotrans-IGBTs circuits. Note the reason to use IGBT is that the switching is so fast that arc quenching need not be done which makes the switching system cheaper overall.

## 12.3 Isolation Transformer

Another transformer is the isolation transformer as shown in Fig. 59. These are used to clean noisy AC waves. It is simple a 1:1 turn ratio transformer. That is, if there are 100 turns in the primary coil there will be 100 turns in the secondary coil. Isolation transformers can cut out electrical noise (Fig. 60), spikes or surges (Fig. 61) and harmonics (Fig. 62 & 63) which flow only on the positive side or only at the negative side (meaning DC harmonics). Harmonics are waves other than the fundamental wave (generated by the power company). TBut recently three groups of home appliance are creating harmonics which do not cancel each other out, causing higher current at the N wire. The three groups of appliances are motor controllers (VFD in air-conditioners, washing machines and refrigerators). SMPS (Switch Mode Power Supply of computers, TV, radios, LED lamps and cell phones chargers) and CFL (Compact Florescent Lamp). All these three groups of appliances work differently but have one thing in common; they rectify the power company's AC to DC and then chops the DC into high frequency AC square-waves. A single VFD motor controller can create up to 25 harmonics. The harmonics are created at the chopping point or the sudden high voltage to zero voltage point as it is switched off by an electronic switch at high to create a square wave. At one moment of time, the magnitude of these harmonics does not add to a low value, making it necessary, as in the USA high-tech data centers to specify the N wire to be 50% bigger in size compared to the phase wires.

Note even a sine wave travelling only on the positive side is DC because the current does not travel backwards or go below the X-axis; this is a "dirty" DC wave which most modern devices like cell phones cannot handle. An SMPS will provide flat enough DC for cell phones and TVs. But the flattest DC is from a battery. Isolation transformers are normal looking transformer but its ratio is 1:1 as shown in Fig. 59. Isolation transformers can clean up to 65% of the defective waves. To achieve higher percentage cleaning of the waves, the equipment in Table 5 can be used. Filters have become important recently because VFD (Variable Frequency Drives), SMPS (Switch Mode Power Supplies) and CFL (Compact Florescent Lights) creates lots of noise and harmonics. It is due to all these harmonics that in Data Centers in USA, the specification for neutral wire is now two times the size of the phase wires. The harmonics, especially the third harmonic which is in phase with the fundamental wave can superimpose with the fundamental wave, causing it to have a much higher amplitude so the addition of the currents in the phases do not add to a small number anymore in the N wire which is why it need to be sized to be two times the phase wire. Note VFD and SMPS are new devices which has become very rampant and the old voltage reducers (transformer rectifiers capacitor circuit) are becoming non-existent. Even up to 1990s radios were made with the old voltage reducers but not anymore. No manufacturer will opt for the old system when the new one is so much cheaper.

The load in an induction motor is a balanced three phase load because all three coils of wire utilize the same current or else the motor will be jerking. Since the three phase currents are 120 degrees out of phase, they cancel out each other leaving zero current in the neutral wire. Note, the neutral wire is the return path of all three-phase wires. In a water pipe system, this return path should be three times the size of each 'phase' pipes. At every moment of time the current in the three phase wires add and minus to result in zero current. So, for three-phase induction motor, N wire is not connected.

In a normal three-phase home however the three phases L1, L2, L3 (R, Y, B) do not use the same current. For an older home, in the 1970s or 1980s for example, at one moment of time, the current can be L1(R) = +5A,

L2 (Y) = -1A and L3 (B) = -2A. L1 (R) =+5A meaning current is moving away from the substation to the load with a magnitude of 5A. L2 (Y) = -1A (meaning current is moving away from the load to the substation with a magnitude of 1A). After the loads L1, L2, L3 (R, Y, B) are joined together using the N wire, this wire consequently goes to the substation N. The current in the N wire where all the phase wires goes through after the load will be +5A-1A-2A = +2A. This is roughly equal to or less than the current in any of the three-phase wires. So, the practice since the beginning of AC power usage till today has been to use the same wire size for all three phases as well as N.

Table 5 is a list of filters to clean out these harmonics. As can be seen isolation transformers clean 65 percent of noise which is cheap and good enough for most installations. This author has heard of installations spending 2400% more than the cost of isolation transformers to clean waves.

The 3$^{rd}$ harmonic is the most concern for power system controllers in factories. The reason is the 3$^{rd}$ harmonic is in phase with the 1$^{st}$ harmonic or the fundamental frequency generated by the power company. Being in phase, it can superimpose and create a higher amplitude waveform. The first few harmonics have high energy and superimposition causes problems but the harmonic number get higher, the energy content is small and superimposition with them do not cause significant problems.

Table 5: Electric power filters available in the market

| | |
|---|---|
| Active Harmonic filters | 95% |
| 18 pulse Drive System | 95% |
| Boradband Blocking Filter- Drives | 92% |
| Neutral Blocking Filter | 90% |
| Neutral Cancellation Transformer | 90% |
| Passive Harmonic Filter | 85% |
| Harmonic Mitigating Transformers | 85% |
| AC line reactors | 65% |
| DC reactors for drives | 65% |
| Isolation transformers | 65% |
| Neutral oversizing | |

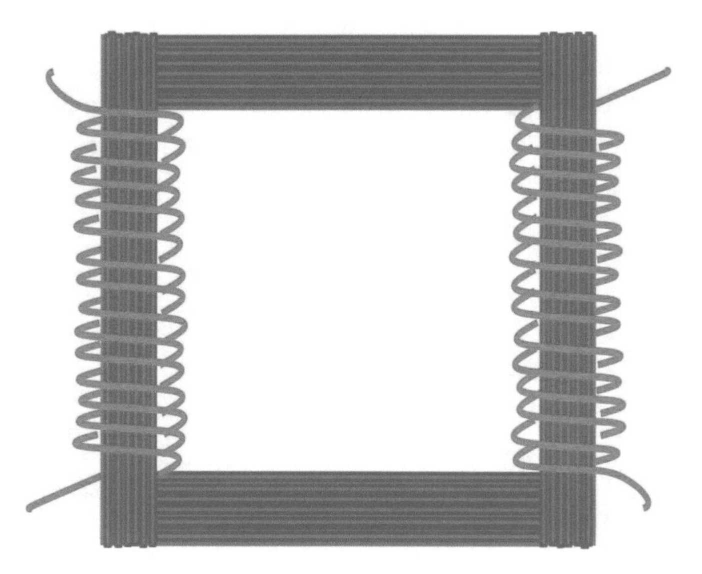

Fig. 59: Isolation transformer same number of turns in primary and secondary

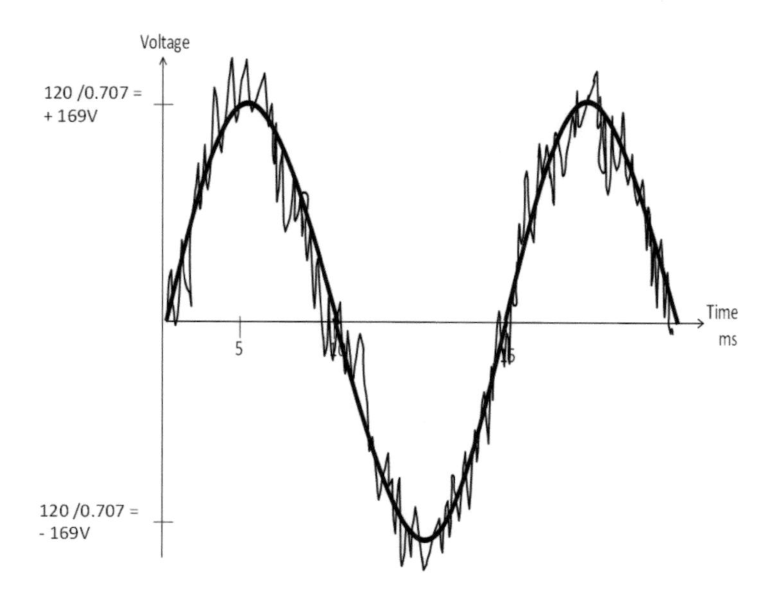


Fig. 60: Noise

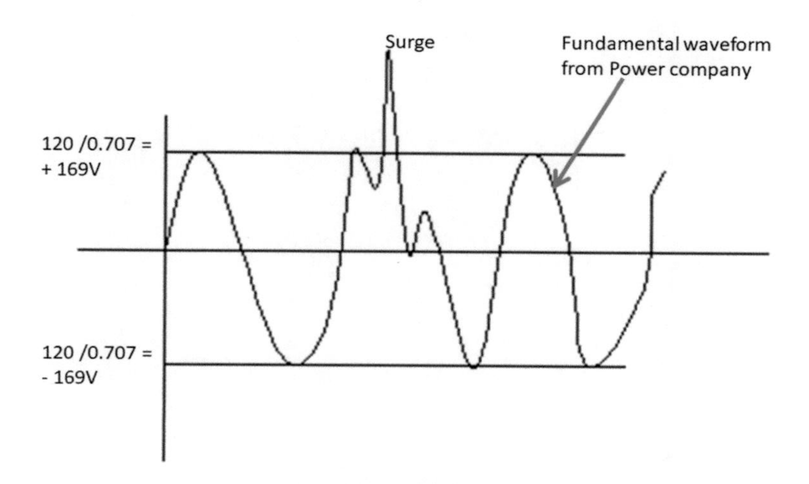


Fig. 61: Surge or Spike

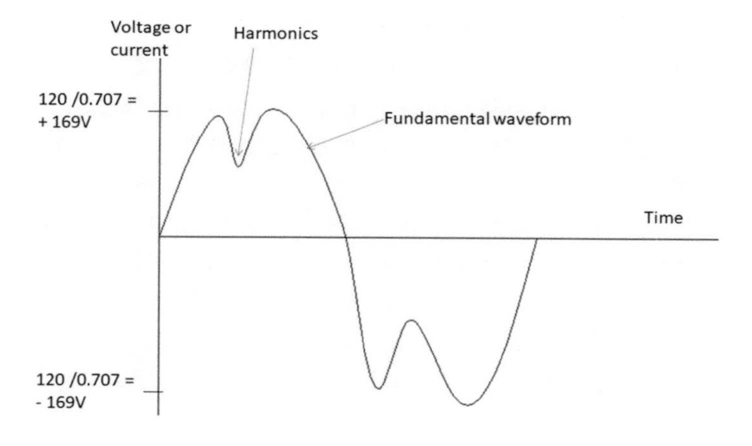


Fig. 62: Harmonics

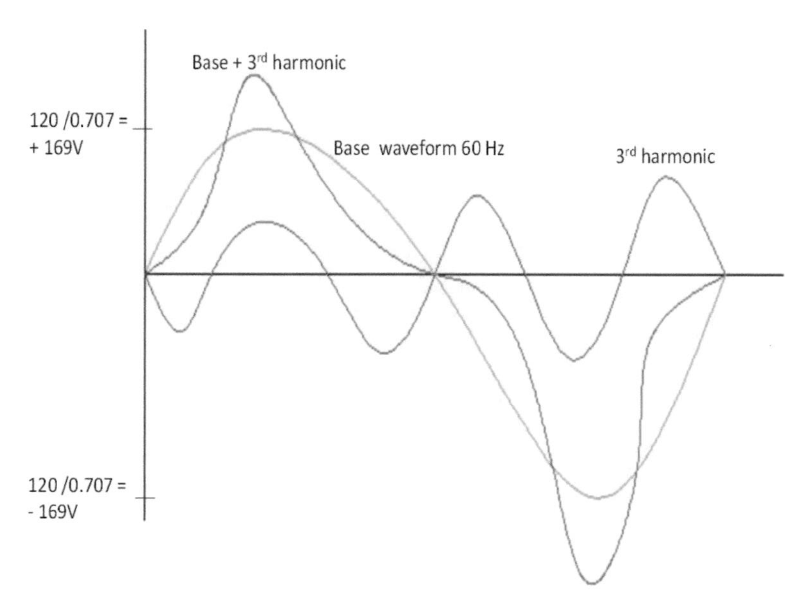


Fig. 63: Harmonics, especially 3rd harmonic must be got rid of

## 12.4 Other transformer developments

Modern scientists are using the transformer principle to replicate Tesla's wireless electric power transmission. At low frequency of 50 to 60 Hz, the Primary and Secondary coils are separated by a paper and an iron core is necessary to conduct the magnetic field but if the frequency is increased to very high values like megahertz, electric power can travel through air without an iron core. Scientist from MIT (Massachusetts Institute of Technology) has achieved a coil in front of a classroom transmitting electric power to a classroom of people each having a laptop with a receiving coil. Comparing this with Tesla in 1920s transmitting wireless power to the ionosphere 1000 km away indicates that Tesla's achievement in 1920 is more advanced than the top engineering university in the world today. Today wireless charging has become common in powering cell phones because they can save time of plugging in the cell phone and thereby damaging the power/data port of the cell phone.

The transformer principle is used to operate radio waves. At 60Hz (or 50Hz) an iron core is needed to transmit AC field lines from the Primary coil to the Secondary coil. But at 3 kHz to 300 GHz frequencies of radio waves transmission is through the air over long distances. Both Amplitude Modulation (AM) and Frequency

Modulation (FM) are used in radio. In AM, songs are carried on carrier waves like 690KHz (range is 530kHz to 1700kHz). The 690kHz carrier wave is an AC sine wave and the songs are carried on top of it (multiplexed onto it). A loud portion of the song is represented by a high amplitude on top of the carrier wave and a soft portion of the song is a low amplitude wave on top of the carrier. In FM station like 97.7MHz (range is 88 to 108 MHz) a loud portion of the song is a higher frequency wave but fixed amplitude multiplexed on top of the 97.7MHz carrier wave and a soft portion is lower frequency but same amplitude wave multiplexed on top of the 97.7 MHz carrier wave.

# Chapter 13

# Electrical formulas

As shown in the middle of Fig. 64, drawing two triangles and putting $P_{VA} = VI$ first then the side of the triangle adjacent to the angle is $VI\cos\theta$ and the side of the triangle opposite the angle is $VI\sin\theta$. For three-phase triangle add $\sqrt{3}$ to the $P=VI$ formula giving $P=\sqrt{3}VI$ but note that the voltage is V=208V and not V=120V (V=415V and not V=240V British) in the single-phase

Fig. 65 shows the power triangle. Various universities and countries use various terms for the three forms of power and they are all represented in the triangle of Fig. 65. The power factor (PF) is the cosine of the angle $\theta$.

formula. Then the side adjacent to the angle is $P=\sqrt{3}VI\cos\theta$ and the side opposite the angle is $P=VI\sin\theta$. With this method, six formulas can be remembered using one formula ($P=VI$ for single-phase resultant power) and two triangles. A normal human being will find it hard to memorize all six formulas (including this author) without the use of the two triangles. And all these six formulas must be remembered because all are well utilized in the electrical power industry.

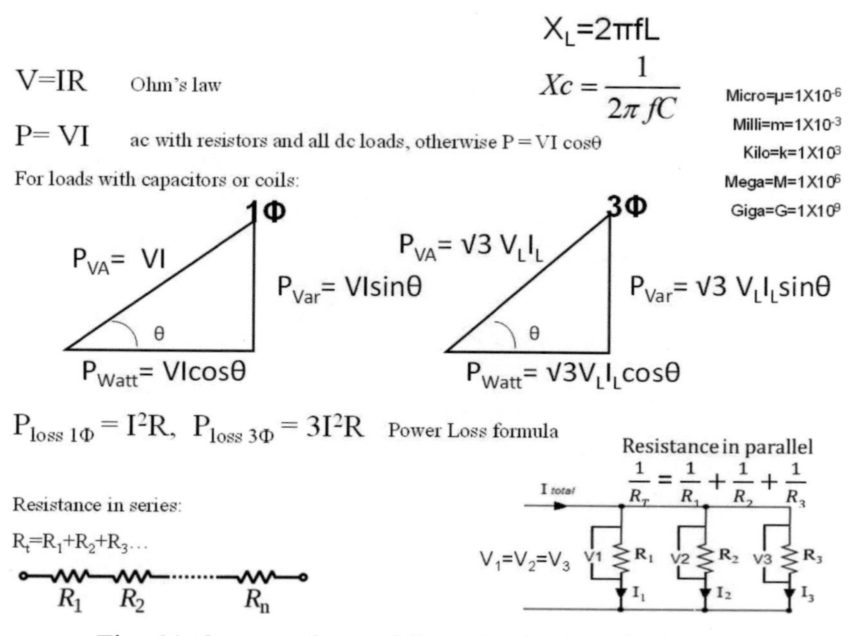


Fig. 64: Commonly used formulas in electrical power

# PF = cos θ

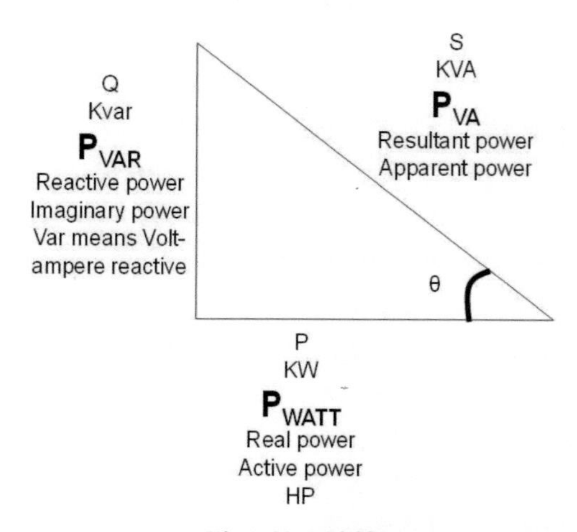


Fig. 65: Different types of power

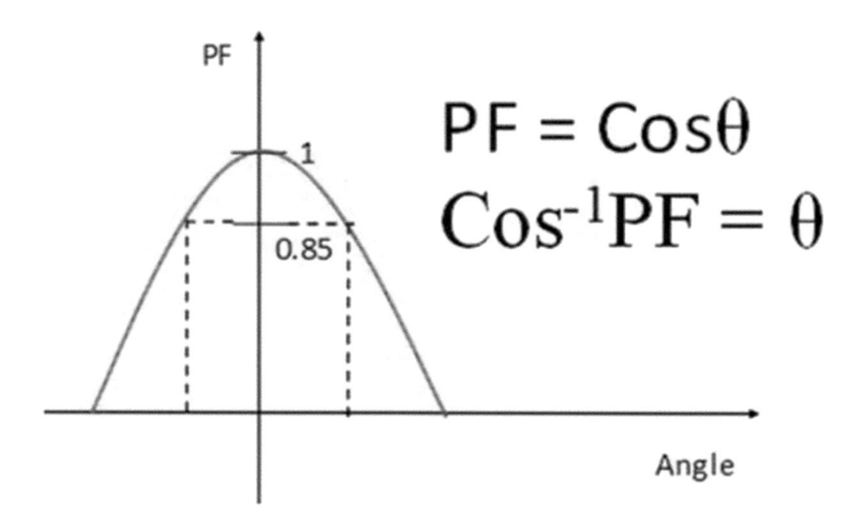


Fig. 66: Power Factor curve

PF is a measure of how much the current waveform shifts (phase shift) from the voltage waveform. Power companies always want no phase shift at all, so big factories are forced to compensate this phase shift of V versus I with the use of capacitors and PF regulators. PF regulator is the brain that determines how many capacitors to switch on in accordance to the number of coils energized. In the old days, not much phase shifting occurred in homes because much of the load was resistive (Edison bulbs, water heaters etc.). Note in a purely resistive load, there is zero shifting of V and I. V and I can only shift of AC power goes through a coil or a capacitor. Of course, there is no shifting in DC which has no waveform. Today, as humans use more coils as in fans, air-conditioners and florescent tube ballasts, the shifting is quite bad. In other words, the PF of homes are quite low. Also, in the old days, because most of the loads in homes were resistive (like Edison bulbs), power measuring meter was designed to measure kWh (real power) but as humans changed their lives styles (using more coiled fans, ballast in florescent tubes etc), power companies are increasing losing money because they supply kVA but bill customers kWh. The hypotenuse (kVA) of the power triangle is longer than the adjacent to angle side (kWh). In many countries, new meters are installed that measures kVA. In those homes, people have experienced their electricity bill going up by up to 30% (hypotenuse is longer than adjacent side by about 30%), given the very low PF of today's homes with lots of coiled loads.

The best way to understand the three types of power is by studying a three-phase induction motor carrying 5kg load. The work to move up the 5 kg against gravity is the real power in watts. While the motor carries the 5 kg, power is expended to create magnetic field lines in the stator coils of the motor. This power is called the reactive power in vars. And while this motor is carrying the 5kg mass the power company is supplying resultant power in VA units.

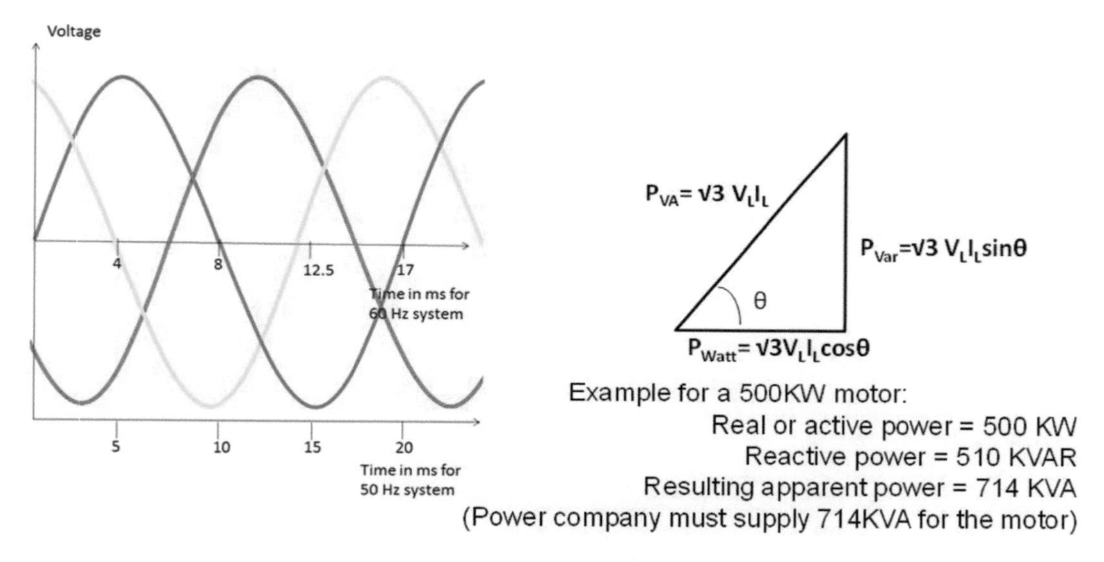


Fig. 67: Further detail of motor power consumption

Another real and measured example is the motor depicted by the triangle of Fig. 67. The load carried is 500kW, the coil's magnetic field strength at that moment utilizes a power of 510 kvar and the power company is supplying at that moment 710 kW. It seems logical that if two power is used, one to carry the load and the other to energize the coils, they should be added to get the total power but because the two power are not in the same direction (their vectors are different), Pythagoras theorem should be used to get the resultant power consumption which is the hypotenuse.

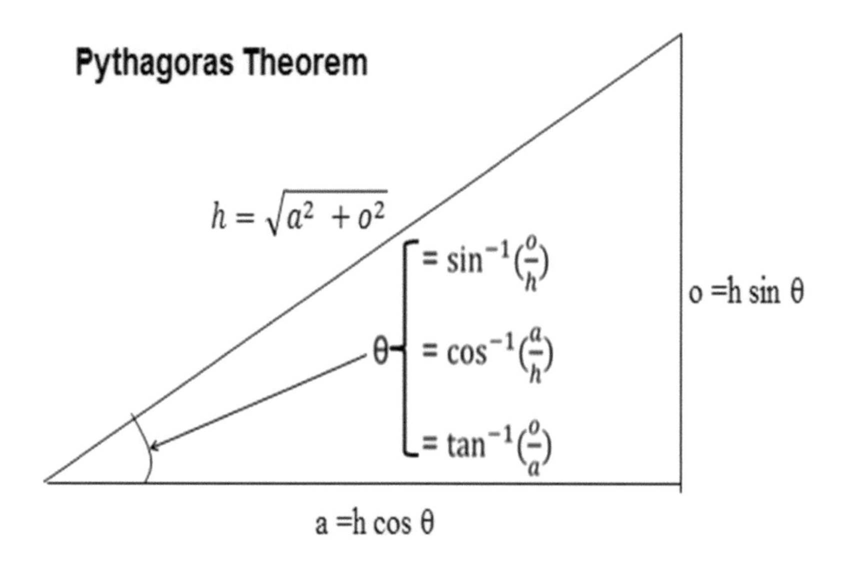


Fig. 68: Pythagoras Theorem

The Pythagoras Theorem shown in Fig. 68 is the most utilized formulas in engineering. Civil engineers can be seen on as we drive on roads looking through a theodolite while his colleague holds up a long ruler. He then needs to use Pythagoras theorem and do triangulation to find heights of various spots. This was the main method to find elevation of mountains before satellites and GPS (Global Positioning System) became the norm. Mechanical engineers use Pythagoras theorem to calculate carrying capacities of cranes which are shaped like triangles. And electrical engineers use Pythagoras theorem to calculate the three types of power, real, reactive, resultant and angle in-between them, in the form of power factor (PF) which is the cosine of the angle between real and resultant power.

In the navy, if a ship has a friendly ship at say 20 km away which can be seen with a theodolite and an enemy ship is at an unknown distance away but can be seen with a theodolite; naval personnel will first point the theodolite at the friendly ship and know how far this ship is from his ship. He then uses the theodolite to view the enemy ship location. Now he has the angle between the two ships and the distance to the friendly ship. So, he can calculate the distance the enemy ship is from his ship and shoot it with a bomb etc. In Pythagoras theorem, there are three lengths of sides of the triangle and the angle is between the adjacent (adjacent to the angle) and the hypotenuse. That is four numbers. If one knows any two of these four, one can calculate the rest.

## 13.1 Power Factor

The biggest problem with the usage of AC electric power is the fact that the voltage and current tend to flow at different speeds when moving through an inductor or capacitor; it does not happen when AC flows through resistive loads. It does not happen in DC also. Since electric power lines are capacitive (underground cable are more capacitive) and inductive (overhead lines are more inductive) this problem must be rectified especially after the electric power has travelled very long distances. This degree at which they flow at different speed is measured by the angular distance between them, θ and is generally represented as cosine θ or PF (Power Factor). This process of getting the voltage and current to flow at the same speed is called compensation.

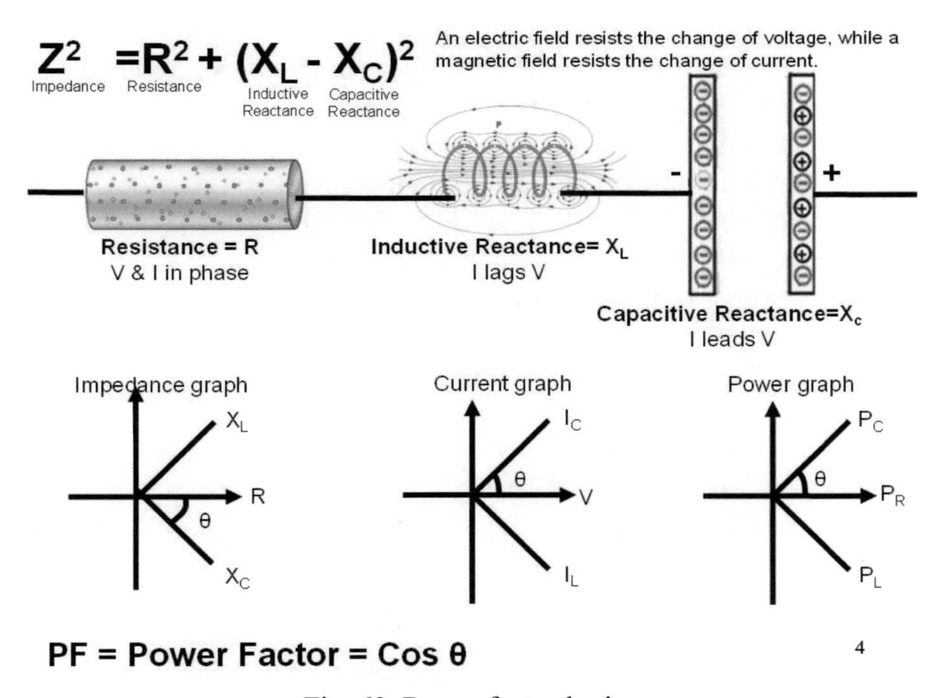


Fig. 69: Power factor logic

Fig. 69 shows electrons within copper colliding with the ions (atoms with less electrons since conductors love to give off electrons and insulators love to take in extra electrons) thereby slowing them down; this is termed resistance (R). The property of conductor atoms to release electrons results in the existence of free electrons within them. If a voltage (Electromotive Force = EMF) is applied to the two ends of a piece of wire, the free electrons will flow as current. Note the electrons do not flow from one end to the other even in DC.

An electron will collide with the next one and this will hit the next one and so forth till the end of the wire and this is termed current. The higher the voltage, the higher the current flow and thereby the higher the power as P=VI. Note that current flow (flow of holes) is in the opposite direction of electron flow and there can be no current flow without electron flow. In other words, current flow is exactly proportional to electron flow but in the opposite direction; we are just living with the mistake that Benjamin Franklin made. For an insulator, even if an EMF is applied on two ends of the wire, the electrons are not free to move so there is no current flow. The more the ions within the conductor vibrate; as when it is heated, the slower electrons move through it. This is a general idea that many people have not realized, if lots of insulation is put around conductors as in an underground cable, the hotter the conductor will be. Try being in an enclosed room without air-conditioning. Actually, underground cable has another problem because its structure is that of a capacitor so in addition to resistance due to ions vibrating because of heat, capacitive reactance slows it further. This capacitor formed in an underground (U/G) cable has to be energized two times every 20ms in single-phase U/G cable because the peak voltage happens two times every 20ms. In a three-phase U/G cable the peak voltage happens six times every 20ms so the capacitor in a three-phase U/G cable will need to be charged six times every 20ms. The old wiring in homes was wires clipped on walls. This was the cool for the wires. These days, the wires are run through plastic conduits thereby increasing the heat around conductors. The latest idea is even worse; burying the wires with conduits within walls, making them even hotter; try staying in an enclosed room without ventilation. As an indication of the current carrying capacity of plastic covered cables versus bare cables the following statistics are useful:

1) The longest undersea cable (heavily covered by plastic) so far, is the **580 km,** 450kV, **700MW** XLPE HVDC cable between Norway and Netherlands.
2) The longest bare overhead line is the **2,385 km,** 600kV, **7.1GW** HVDC Rio Madeira transmission link in Brazil.
3) A close second is the **2,090 km, 800kV**, 7.2GW HVDC Jinping-Sunan transmission link in China.

**Therefore, bare O/H cables can carry 10 X more power 4X further.** Cables with insulation surrounding them are far less capable than bare overhead cables. Added to the above statistics, underground cables fail much more often than overhead cables and the overall cost of underground cables is up to 400% higher than bare overhead cables.

In Fig. 67, as soon as a voltage is applied across a coil of wire, a magnetic field builds up around it. There will be a N (when electrons enter it in a clockwise direction) and a S (when electrons enters it in anticlockwise direction). Placing a solenoid below a piece of paper and scattering iron filing on the paper will result in the shape of field lines exactly as seen when a permanent magnet is placed below the paper. Magnetism is experienced by a coil of wire when electrons are moving in each coil of a solenoid in parallel to each other. In a piece of iron, there are domains where unpaired electrons spin parallel to each other within each domain. This is similar to electrons moving in the same direction parallel to each other in a solenoid. So, each domain is a powerful magnet. But the electron spin in the neighboring domains is oriented in a different directions. Just like numerous tiny domain sized magnets all oriented in varying directions. Thus the overall iron is not magnetic because the magnetism in each domain cancel out each other. But if this iron is placed inside a solenoid, all the electron spins in all the domains get aligned in one direction, resulting in a powerful magnet. Otherwise a permanent magnet can be swiped over a piece of steel needle (about 95% iron) about 40 times (as the son of this author did following a science book), the electrons spins in all the domains align, turning the steel needle into a magnet.

So, as a coil or inductor is energized it becomes a magnet which will slow current flow, just as placing a permanent magnet near a current carrying wire will slow current flow. When a voltage is applied to a coil, a magnetic field is immediately created around the coil. Then slowly the current flows into it. In the electric industry this is termed, current lags the voltage. **Thus, current lags the voltage in an inductor.** Eventually after a buildup of the magnetic field around the solenoid, it will break down leaving zero voltage across the inductor and the process of building magnetic field repeats itself.

Fig. 67 shows a capacitor. Initially both plates of the capacitor are not charged. At this moment, if voltmeter probes are applied across the two capacitor plates, the voltage reading will be zero since there is no difference in the potential of the two plates. But when a current is applied to one plate, it gets filled with electrons. Now there is a difference in the two plates. Now, as voltmeter probes are again placed on the two plates, there will be a voltage reading. This is because one plate is not charged but the other is charged with recently filled electrons. Since the electrons must enter before there is a voltage, in the electrical industry this is termed, **in a capacitor, current leads the voltage.** Eventually the buildup of electrons increases to such an extent that there will be an avalanche of electrons to the other plate, and the process repeats itself. Most electrical engineers remember the leading or lagging using capacitors rather than inductors; after remembering how the capacitor works, just remember the inductor has the opposite effect or **in inductor, current lags the voltage**. Some Universities teach students using mnemonics like CIVIL. Going from left to right CIV represents in Capacitor, I leads the V. VIL represents V leads the I in inductors. Some ask why inductor is represented by L. It is simple, current in already represented by I so L is used to represent inductor. Also, L is the first letter of the surname of Heinrich Lenz so, L is used to honor his work in electromagnetism. In the opinion of this author using mnemonics is dangerous because of the tendency to translate the alphabets wrongly. Remembering how the capacitor works is the best way to remember which one is leading and which is lagging. Current or electrons has to go in one plate before there is a voltage difference between the two places thus current leads the voltage. Note the voltage is always on the X-axis and the current line is always drawn above (leading) or below (lagging) the X-axis. For remembrance, it can be taken like an anticlockwise race. If I line is above the horizontal V, I has won the anticlockwise race, thus I leads the V.

In industries, capacitors are used to balance the inductive machines which do all the work; these can be motors which have lots of coils. These capacitors are placed in the capacitor bank of the MSB (Main Switch Board) as showed in Fig. 72. Inductive machines like motors, ballast of florescent tubes or induction cookers causes the current to lag the voltage. Capacitor bank units automatically energize the appropriate number of capacitors according to inductive loads used. This switching on of capacitors will cause the current to lead the voltage. So, the lagging of current caused by the motor is compensated by the leading of current caused by the capacitors. This will bring the V wave and I wave to be in phase. The ‘brain’ that switches on or off capacitors is the Power Factor Regulator as shown in Fig. 71. Fig. 73 shows a closer view of the capacitor used in MSB.

Table 6: Effects of PF on a 90KW motor

| Power Factor | 1 | 0.9 | 0.8 | 0.7 | 0.6 | 0.5 |
|---|---|---|---|---|---|---|
| Load (kW) | 90 | 90 | 90 | 90 | 90 | 90 |
| Current (A) | 187 | 208 | 234 | 268 | 312 | 375 |

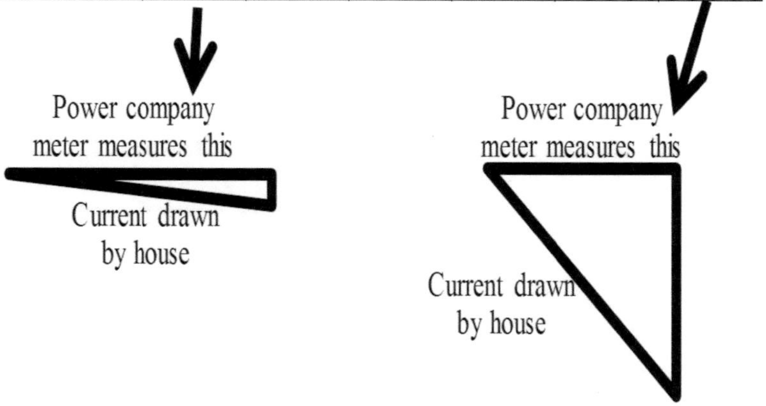

Power company meter measures this
Current drawn by house

Power company meter measures this
Current drawn by house

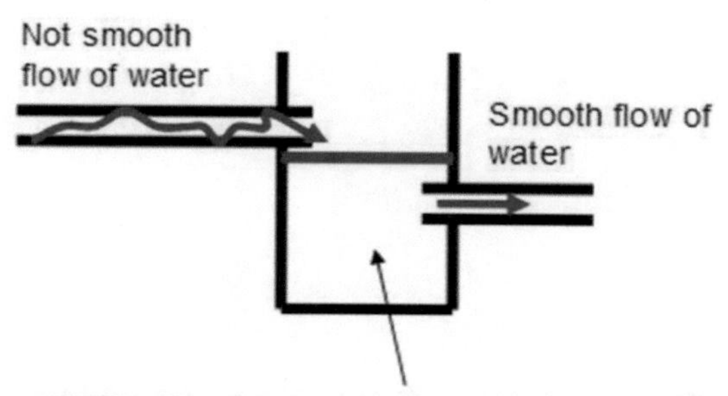


**This water tank is equivalent to a capacitor**

Fig. 70: Capacitor effects

**PF Regulator in MSB**

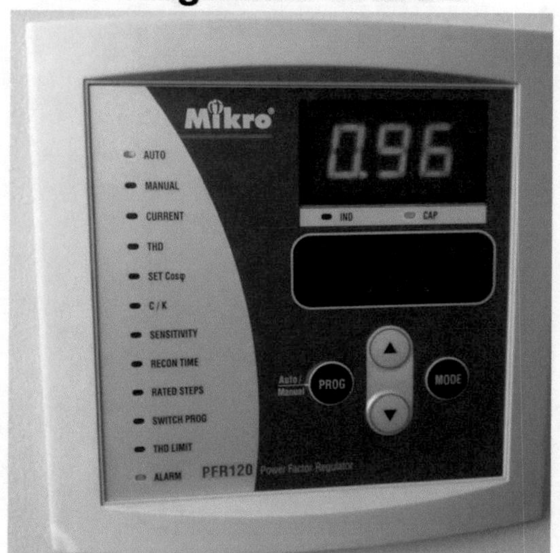


Fig. 71: PF regulator of a MSB

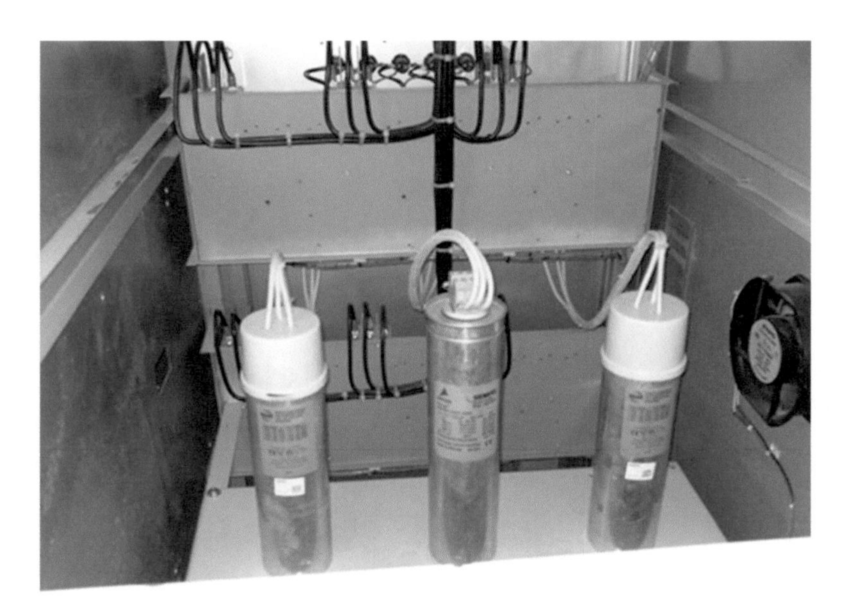

Fig. 72: Capacitor bank

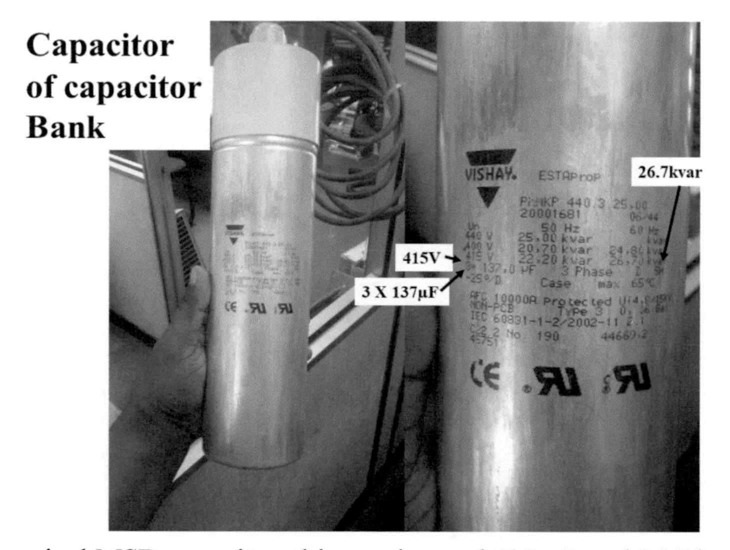


Fig. 73: A typical MSB capacitor, this one is rated 137 μF and 26.7 kvar at 415V

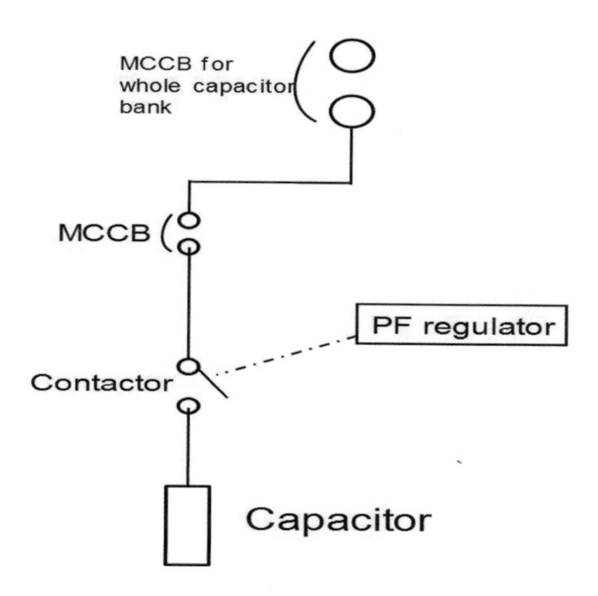


Fig. 74: Capacitor bank single line diagram

Fig. 74 is the schematic for the PF regulator controlling the capacitors in a typical MSB.The schematic for the PF regulator controlling the capacitors is as follows:

As mentioned previously PF is the cosine of the angle between the I waveform and the V waveform. In industry the I wave and the V wave starts to flow at different speed if the load is a motor (a coil) or the ballast of a florescent tube (an autotran). Two methods used to bring back the V and I to the same speed is by use of a capacitor bank which are located in all factories (by regulation), in all 33kV substations and some EHV substations or via playing around slightly with the DC going into the rotors of a large hydroelectric generators that is not connected to its prime mover (the falling water). Power companies call this running the generator in synchro mode, that is running the generator as a synchronous motor. This is always done in a hydroelectric power station because here the generator can easily be run as motor since it is not fastened to any prime mover as in a gas turbine generator, ICE (Internal Combustion Engine) generator or coal fired steam engine generator. A recent third method of controlling the PF is by using a SVC (Static Var Compensator). This is mostly utilized by power companies or large steel mills factories. The SVR has industrial size capacitors and reactors (coils or inductors shown in Fig.104) plus a microprocessor to dynamically balance the phase-shift of the I with respect to V. Reactors or any coils have the property of getting rid of harmonics so other than PF compensation SVC can also remove harmonics in the grid.

Phase voltage is the voltage between any phase and neutral; while line voltage is the voltage between two phases. In overhead lines the schematic in Fig. 75.

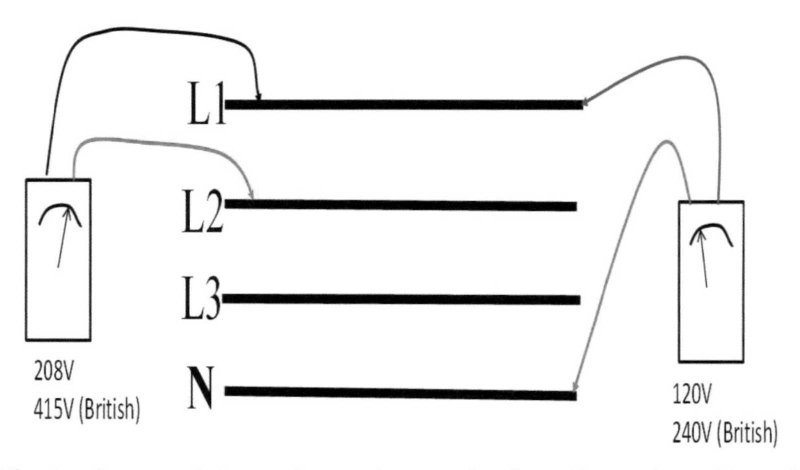


Fig. 75: Single phase and three-phase: the standard configuration of housing estate overhead lines. L1, L2, L3, S/W, N (R, Y, B, S/W, N) from top to down.

# Chapter 14

# Calculations

A sample calculation is as follows:

1) What is the power loss when phase voltage is connected to a 150 Ω resistor?

Answer:

$$V = IR \quad (52)$$

$$240V = I\,(150)$$

$$\frac{240}{150} = I$$

$$1.6A = I$$

$$P_{loss} = I^2 R \quad (53)$$

$$P_{loss} = 1.6^2(150)$$

$$P_{loss} = \underline{\underline{384\,Watt}}$$

2) What is the total resistance of a 20 Ω a 50 Ω and a 100Ω resistor in parallel.

$$\frac{1}{R_T} = \frac{1}{R_1} + \frac{1}{R_2} + \frac{1}{R_3} \quad (54)$$

$$\frac{1}{R_T} = \frac{1}{20} + \frac{1}{50} + \frac{1}{100} = 12.5\Omega$$

In a calculator just type the following to get the answer:

1÷20 + 1÷50 + 1÷100 = 1÷ ANS

Where ANS is the calculator key for the previous answer.

3) What is the current flowing through a 30mH inductor connected to phase supply?

$$X_L = 2\pi fL \quad (55)$$

$$= 2\pi(50Hz)(30mH)$$

$$= 7.42\Omega$$

$$V = IR \quad (56)$$

$$V = IX_L \quad (57)$$

$$240V = I(7.42)$$

$$\frac{240}{7.42}A = I$$

$$\underline{\underline{32.3A}} = I$$

4) What is the current flowing through a 90μF capacitor connected to phase voltage?

$$X_C = \frac{1}{2\pi fC} \quad (58)$$

$$= \frac{1}{2\pi(50Hz)(90\mu F)}$$

$$= 35.37\Omega$$

$$V = IR \quad (59)$$

$$V = IX_C \quad (60)$$

$$240V = I(35.37)$$

$$\frac{240}{7.42}A = I$$

$$\underline{\underline{6.79A}} = I$$

5) What is the total power for three factories owned by one owner nearby each other. The first factory uses 50kW at unity power factor. The second factory uses 80kVA at PF=0.6 and the third factory uses 40kVA at PF =0.7 lagging. Also find the total current and KVA.

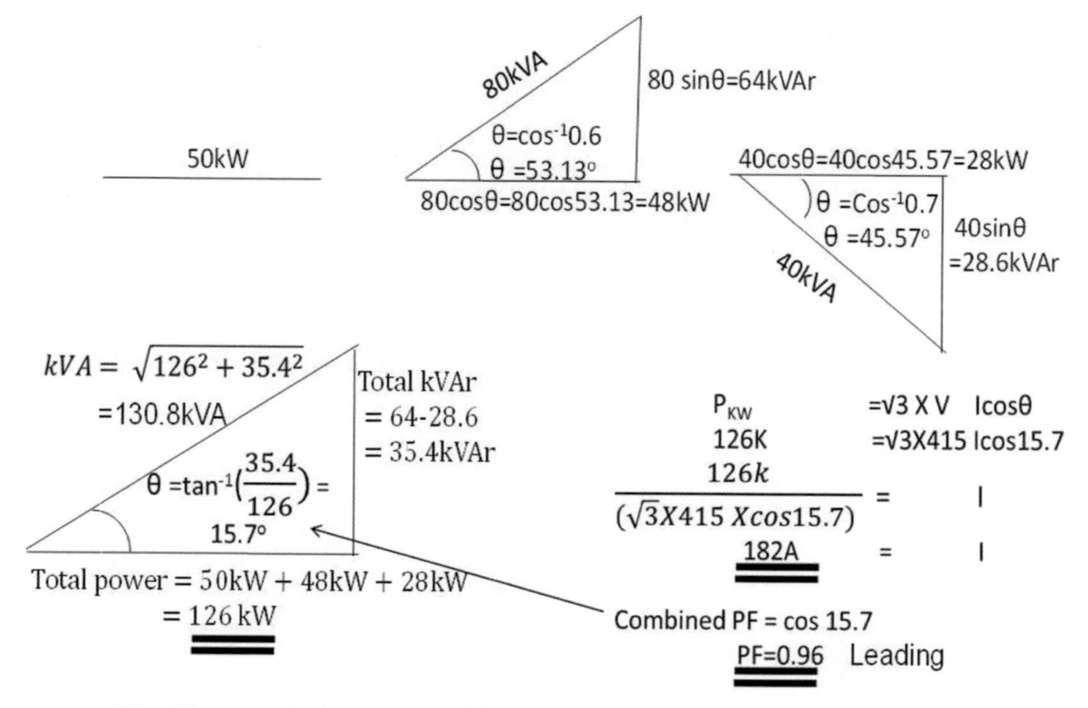


Fig. 76: PF calculation and effects on three buildings with one proprietor

The calculation above depicts a case where three factories are owned by an industrialist. If the power company inspector were to inspect the second 80kVA factory he will report that a heavy fine should be imposed on that factory because the PF = 0.6 while the minimum limit is 0.85 for most countries. Upon inspecting the third 40 kVA factory he will also recommend a heavy fine since the PF = 0.7. But if the engineer hired by the industrialist did the above calculation he can show the power company inspector that if all three factories are taken together the PF = 0.96 (way above the minimum limit of 0.85). So, there should be no fine for this industrialist.

6) The open circuit voltage of a lead acid battery is 12V. When a 25Ω resistor is connected to the terminals of the battery, the terminal voltage (voltmeter probes on the battery terminals) drops to 10V. What is the internal resistance of the battery?

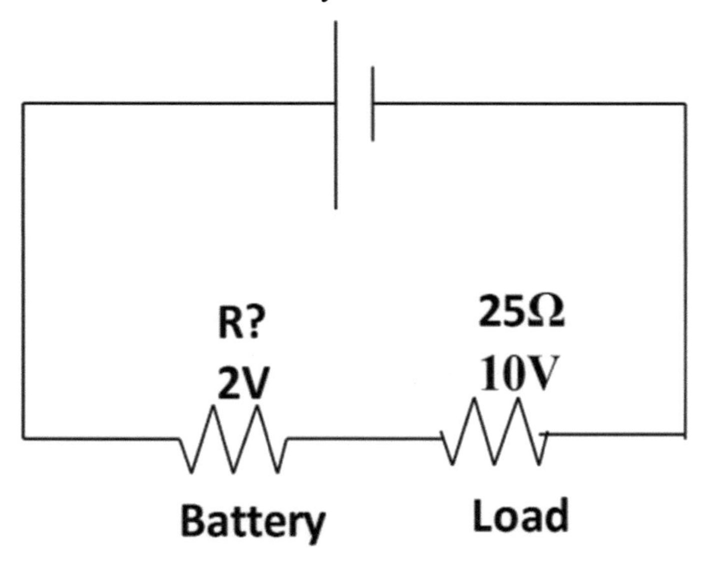


Fig. 77: Battery internal resistance

At the battery internal resistor:

$$V = IR \quad (61)$$

$$\frac{V}{R} = I$$

At the load:

$$V = IR \quad (62)$$

$$\frac{V}{R} = I$$

$$\frac{2}{R} = I \qquad \frac{10}{25} = I$$

$$Therefore\ \frac{2}{R} = \frac{10}{25}$$

$$R\ X 10 = 2\ X\ 25$$

$$R = \frac{2\ X\ 25}{10}$$

$$R = \underline{\underline{5\Omega}}$$

7) Given the grouping factor is 0.55, ambient temperature is 0.65, the circuit is protected by a rewritable fuse rating at 15 ampere, calculate the $I_z$?

$I_z$=the current carrying capacity of the cable. $I_n$= fuse rating. The other factors are described below.

$$I_Z = \frac{I_n}{(C_a + C_g + C_i)} \qquad (63)$$

There are five factors that affect the voltage drop namely:

1) Cable table – the standard cable chooser according to the current flowing in the cable. The American Wire Gauge (AWG) or the IEE (Institute of Electrical Engineers) wire current carrying capacity.
2) Voltage drop – considered if the load is far away (say 100m) away from the source.
3) $C_a$ – a stands for ambient temperature. If the electric installation is in a boiler room, a bigger wire size need to be used and a freezer project can use a smaller wire size. The $C_a$ can be 0.9 in Saudi Arabia and $C_a$ can be 0.5 in the North Pole.
4) $C_g$ – g stands for grouping. For example, this author consulted for a water works project where the architect specified only one conduit for all the cables. So each cable need to be bigger size.
   $C_i$ – i stands for insulation. There were many types of insulation used previously. One of the first was natural rubber the rubber tree. Today cable insulation worldwide has merged to two types, namely PVC (polyvinyl chloride) and XLPE (cross-linked polyethylene).

So for the above question:

$$I_Z = \frac{I_n}{(C_a \, X \, C_g \, X \, C_i)}$$

Where In =15A, $C_a$=0.65, $C_g$=0.55, $C_i$= not given so can use 1

$$I_Z = \frac{15}{(0.65 \, X \, 0.55 \, X \, 1)} = 41.95\text{A}$$

Note the value $I_z$ is the current carrying capacity of the cable at these conditions of environmental temperature, grouping and plastic insulation. If the denominator is smaller like $C_a$=0.1 (as in the North Pole), the current carrying capacity is higher. If the denominator is $C_a$=0.9 (as in Saudi Arabia) the current carrying capacity reduces. If we put 1, the maximum value, the $I_z$ will be the lowest possible value.

8) A balanced 25A is supplied from 415 V Distribution Board (DB) located 150m away, total voltage drop must not exceed 4% supply voltage:

   a) Calculate max permissible V-drop
   b) Choose most suitable cable size from table given below:

| Cable size (Sq mm) | 6 | 10 | 16 | 25 |
|---|---|---|---|---|
| Current rating (A) | 34 | 46 | 62 | 80 |
| Voltage drop per Amp per meter (MV) | 6.4 | 3.8 | 2.4 | 1.5 |

The word balance indicates an induction motor which is the only load where the three phases use the same current. The maximum possible voltage drop is 415 X $\frac{4}{100} = \underline{\underline{16.6\text{V}}}$

The voltage drop equation is:

$$V_D = \frac{(MV)IL}{1000} \quad (64)$$

We can try 6mm$^2$ cable because it can carry 34A which is higher than 25A used by the load.

$$V_D = \frac{(MV)IL}{1000}$$

$$V_D = \frac{(6.4)\,X\,25\,X\,150}{1000}$$

$$V_D = 24\text{V}$$

So 6mm$^2$ cable cannot be used because the $V_D$ is higher than 16.6V. 6mm2 cable has a voltage drop of $\frac{24}{415}X100 = 5.78\%$ which is above the 4% limit.

Try 10mm2 cable

$$V_D = \frac{(MV)IL}{1000}$$

$$V_D = \frac{(3.8)\,X\,25\,X\,150}{1000}$$

$$V_D = 14.25\text{V}$$

This cable is alright because it's voltage drop is less than 16.6V. The voltage drop is actually $\frac{14.25}{415}X100 = 3.43\%$, less than the 4% limit. <u>So, it can be used for the 150m distance away load.</u>

# Chapter 15

# Magnetism

In the early days, magnetism was thought to be an innate nature of chunks of iron ore struck by lightning. The early Chinese would walk to the top of an iron ore filled mountain to search for iron ore rocks that was struck by lightning which would be magnetic. They used these rocks to swipe across a non-magnetic iron to make more magnets.

Then in AD1040 Wu Ching Tsung Yao wrote in the 'Compendium of Military Technology' that compass needles could also be made by heating a thin piece of iron in a container, often in the shape of a fish, to a temperature above the Curie Point which is the temperature at which a magnet will lose its magnetism, then cooling it in line with the earth's magnetic field. This ended the need to wait for lightning to strike a mountain of iron.

In 1821, Hans Christian Oersted showed that a piece of wire carrying current could also deflect a compass, just as a piece of magnet would.

In 1820 Andre-Marie Ampere theorized that magnetism was caused by electric currents and that the same process of magnetism in a permanent magnet occurs in a coil of wire or solenoid.

In 1821, Michael Faraday built the first motor as a piece of DC current carrying wire (positive) hung as a hook on an incoming wire and at the bottom end of this hook was a glass of conductive mercury where the battery's negative terminal was connected. A permanent magnet was placed at the center of the glass. The suspended current carrying wire spun around the permanent magnet, in-effect creating the first motor. Today the liquid mercury and the free hook on the other wire are replaced in a DC motor by two bearings holding the shaft and a set of carbon brushes providing conduction and freedom of movement as the mercury and free hook did.

Later Faraday immediately theorized that the system should be able to work the other way round. He therefore developed the world's first generator where a permanent magnet he moved with his hands generated electricity in a surrounding coil of copper wire. He could measure the voltage of this generated electricity with a voltmeter collected to the ends of the coil. Thus, Faraday invented the motor, the generator and transformer causing most literature to call him the Father of Electricity.

In ferromagnetic materials, the spin of unpaired electrons in small regions called domains makes each of these regions a tiny magnet. But in neighboring domains the spins of unpaired electrons are oriented in a different direction thus overall, iron is not magnetic. The magnetic force in each domain of non-magnetic iron is quite strong but it is cancelled out by other domains where the electron spins are oriented in a different direction. This is just like lots of tiny magnets oriented in different directions; resulting in zero net magnetism on a piece of iron.

Electrons can be equated to tiny magnets so when there is a flow of electrons in a wire, there is an electric field around it. In 1905 Einstein wrote a paper stating that electric fields and magnetic fields are the same phenomenon viewed from a different plane. In a generator, a coil of wire is spun in the rotor and DC is sent to this coil via carbon brushes. This turns it into a magnet (solenoid). In smaller generators, a permanent magnet is used as the rotor. In big generators permanent magnets cannot be used because the heat within the motor is too high and a permanent magnet will turn back into iron (Fe) when it is heated. But the magnetic strength of permanent magnets is increasing over time, mainly due to innovations at Hitachi. So, today increasingly higher power output generators are using permanent magnet rotors. However, the only advantage of using permanent magnet rotors is that there is no need to send DC current to the rotor via carbon brushes so as to turn it into a solenoid. But with the huge increase in carbon bush technology, where even the carbon brush of a good-brand drill (like Hitachi or DeWalt) can last for up to 20 years, the justification for a permanent magnet rotor has decreased.

Fe is magnetic because there are tiny regions within Fe called domains where the free electrons flow in the same direction and therefore parallel to each other. Thereby each of these regions or domains are tiny magnets. In Fe, the orientation of these domain-sized tiny magnets is different, so Fe is not magnetic. But when a permanent magnet is swiped over an iron needle for 40 times, the direction of the magnet in all the domain aligns turning the Fe into a magnet. In a solenoid, the electrons are moving parallel to each other in the coils which is similar to electrons spinning parallel to each other within each domain. In a permanent magnet all the domains' magnetism aligns. This is why a permanent magnet does the same attraction for Fe as a solenoid. The advantage of a solenoid within a generator is that the direction of electron flow is fixed and is much less affected by heat than a permanent magnet. Heat causes aligned domains in magnetic Fe to by unaligned therefore turning it back to Fe.

The free electrons electron spinning parallel to each other within each domains of Fe makes each domain a tiny magnet in one direction. Some are asking why isn't there one particular direction to which there is a slight majority of domain magnets orienting to, resulting in iron attracting another piece of iron a little. The answer to this is analogous to throwing a coin. If we throw 10 times we may not get five heads and five tails, but if you throw it 100 times, you will get closer to 50 heads and 50 tails but if you throw a billion times, the it will be quite surely 50% heads and 50% tails. So, with billions of domains, the likelihood of no one direction having a little more magnetism is surer. When a magnet is used to swipe over a piece of iron, what is happening is that most of the domains get to be oriented in the same direction causing the iron to be a magnet. This is the similar explanation to a LASER light. If you look directly at white light, one sine wave will be along the Y-axis (90 degrees) another sine wave will be at zero degrees another will be at 30 degrees and many different planes. The LASER device makes all these waves in multiple angles to be in one direction becoming a powerful LASER beam that can be made strong enough to cut a thick iron sheet. In a piece of iron, when all domains' magnetism is oriented to be in the same direction, the iron can be a very strong magnet.

The ability of ferromagnetic materials like iron, neodymium, nickel, cobalt, gadolinium and dysprosium to conduct magnetic fields is called relative permeability. When most of the domains of a ferromagnetic material is magnetized in one direction, it will not relax back to zero magnetization, when the inducing electromagnet is removed. The amount of magnetization it retains is called its remanence (similar word to remember). To drive it back to zero magnetism, a field in the opposite direction must be applied above it. The amount of reverse driving field required to demagnetize the magnet is called its coercivity as shown in Fig. 78.

It is the property of remanence

that enables the hard disk industry to store information of disks. To get ever higher storage capacity hard disk, the head is flown lower and lower over the disk. The fly-height is like 0.2 micro inch. Lower fly-heights enables the magnets to be smaller or the ones and zeros to be more numerous, thereby increasing the storage capacity of the hard disk. The head is actually flying, it cannot be suspended at such low heights. The head has wings and spoilers to enable flying at exactly this fly-height like an airplane. To enable writing onto the disk, it holds a coil of wire where DC is continually switched direction to code particular points on the hard disk to be either NS or SN. A NN junction will be a one, a SS junction will also be a one but a NS or SN junction will be a zero. When the head is reading data on the hard disk, the same coil will read and after every eight bits it reads, there will be stop bits. Each eight bits is converted to a character, a number, a color or an instruction, according to a table called ASCII (American Standard Code for Information Interchange).

This seem rather complicated but competitors to this system have been devised like the flash memory we use in pen-drives but in cost-per-byte, the hard disk has no competition. This author worked for an electric utility and helped someone to go for an open-interview for Western Digital (WD) but ended up working for worked for WD for 14 years. At the beginning of working in the industry, there were wild predictions that the industry will close down due to optical compact disks (CDs). But their slow speed in data extraction proved to be a deterrent. Then flash memory was the next thing which was supposed to destroy the hard disk industry but their cost-per-byte was way too high and it has a very much shorted life-span. In hard disk, the magnet can theoretically be flipped forever but NPN junctions cannot survive being bombarded with electrons too many times. Hard disks are damaged if they are moved, causing the head to hit onto the disk, making a ripple on the disk (like when you throw a stone in a pool of water) which will destroy lots of data. This must be why when the worldwide population moved from desktops to laptops, many top business leaders predicted that the hard disk industry days were numbered. But because the public started using clouds to store most of their data, hard disk made a come-back. Hard disks are obviously the winner in Data Centers since cost and durability are critical concerns here. In Data Centers the environmental conditions for the hard disk can be optimized with zero vibration and cool temperatures; note magnets on hard disks cannot withstand high temperatures. A note to readers is you should avoid shaking a laptop as far as possible and always use a cooler to cool your laptop. This author has a 11-year old laptop that is still working due to minimal vibration. There are sensors placed in hard disks of today to detect vibrations and laptop falls which will quickly move the head away from the disk but why take a chance on the hard disk of your laptop.

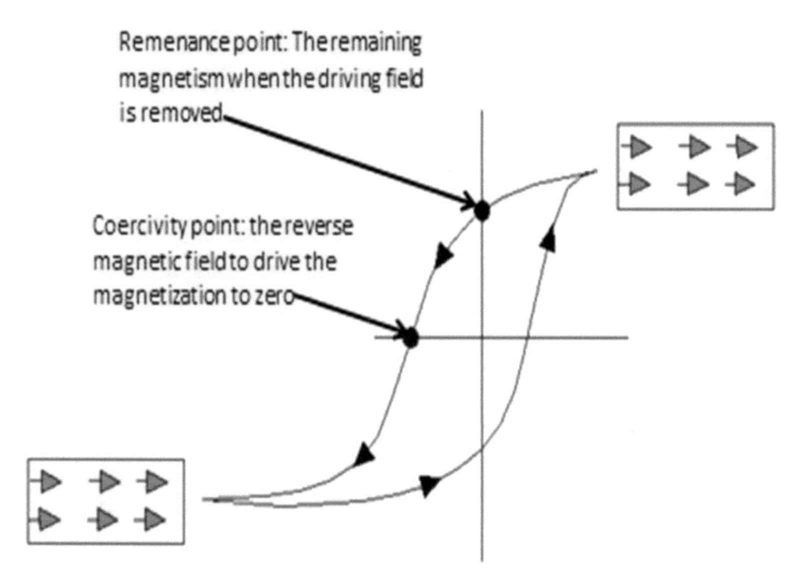


Fig. 78: The Hysteresis Loop

The most powerful permanent magnet uses neodymium. They are used in microphones, professional loudspeakers, in-ear headphones, computer hard disks, wind turbines and hybrid vehicles. 95% of neodymium and other rare earth elements are from China and Hitachi holds more than 600 patents covering Neodymium magnets.

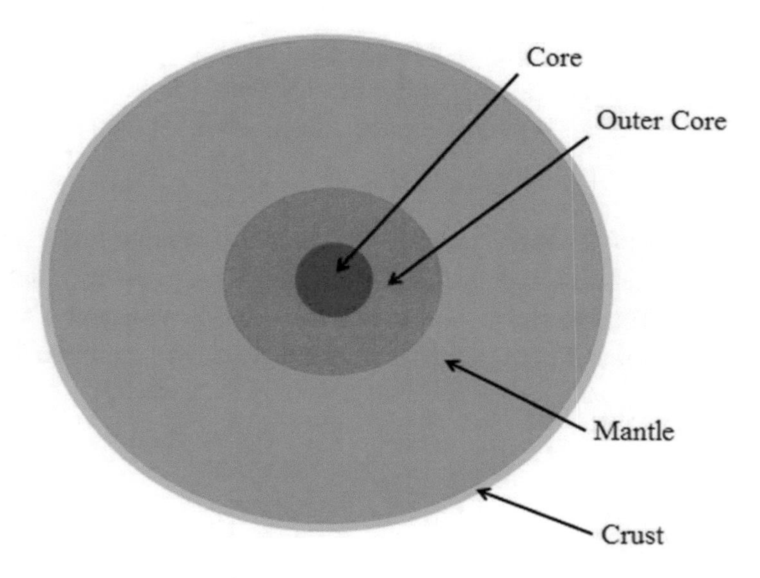


Fig. 79: Magnetism of the earth

The unit used to measure magnetic field is Tesla. The magnetic field strength of the earth is around 0.0001 T. The earth's magnetic field causes a compass placed anywhere on earth to turn such that the north pole of the magnet points to the North Pole of the earth. Note that North Pole is capital letter to indicate that it is the name of a place on earth. Actually, North pole is a south pole of the earth magnet which attracts the north pole of all magnets. The magnetism of earth is caused by the molten iron circulating in the outer core of the earth's crust as shown as the second layer from inside in Fig. 79. The inner core is solid iron despite the very high temperature due to the extreme high pressure here. The outer core is molten iron and because iron atom loves to throw off two electrons, effectively the outer core is a layer of spinning electrons which spins around an axis of the North to South pole. These electrons are spinning parallel to each just as in a solenoid; creating a magnetic field, effectively making the earth a magnet with N and S poles. The earth's magnetic field has a great utility in preventing harmful solar and other rays and asteroids from reaching the earth. The reason why iron is found in the inner and outer core is explained like this: If one pours a bit of granular sugar and milk power in a glass and stir it, the heavier granular sugar will converge to the center of the glass; therefore, when the earth was just a spinning ball of gas, the heavier iron converged to the center and lighter silicon (sand) moved out to the Mantle and crust above.

The theory of how the moon was formed is that an asteroid crashed onto the earth almost totally destroying earth. In this humongous explosion, a huge cloud of silicon dust ($SiO_2$) emanated from the Mantle and crust above it, surrounded the earth and eventually this silicon gathered together to become the moon. Thus, the moon is mostly silicon and therefore has zero magnetic fields around it to prevent harmful rays and asteroids from hammering it daily.

All materials can be classed under three forms of magnetism, namely ferromagnetic, paramagnetic and diamagnetic. Ferromagnetic materials like Fe, Nd, Co, Gd and Dy are the ones strongly attracted to a magnet. Paramagnetic materials: like Al, Mg and $O_2$ are weakly attracted to a magnet. Diamagnetic materials like C and $H_2O$ and all substances which do not fall in the ferromagnetic or paramagnetic categories.

# Chapter 16

## Low Voltage wiring

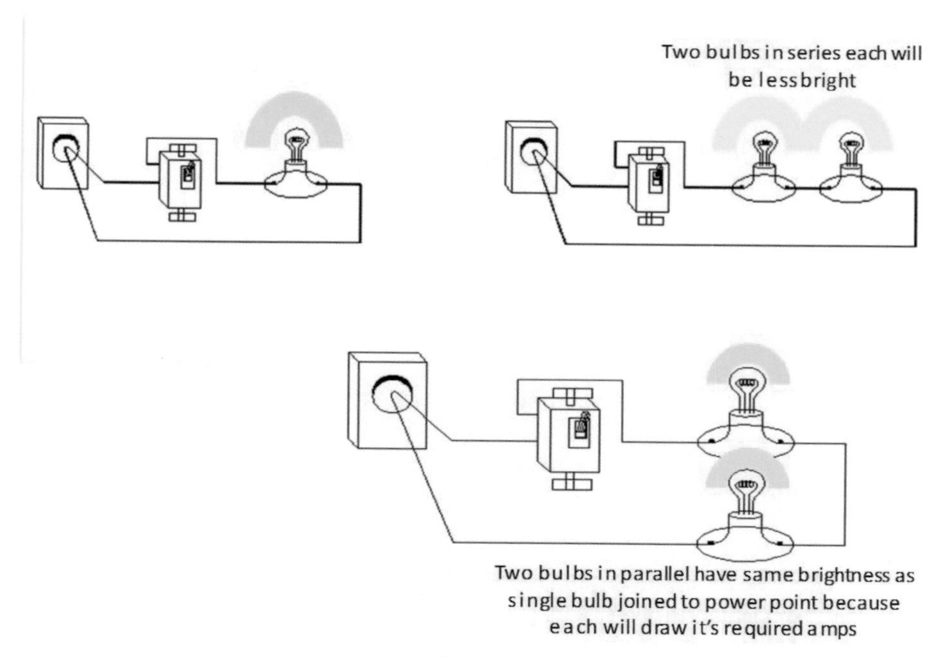


Fig. 80: Home lighting

An experiment was done using two Edison bulbs and the results are in Table 7. Note, current is the same in series and voltage is the same in parallel. Fig. 81 is the standard way home wiring is done. Another experiment was done with three Edison bulbs and the results are shown in Table 8. The result in Table 8 and Fig. 82 and Fig. 83 depicts running Edison bulbs without the N wire. Phase to Phase voltage is injected. An interesting observation is that an Edison bulb is able to withstand a large variation in voltage without being blown, it can withstand voltage from 0V-415V empirically.

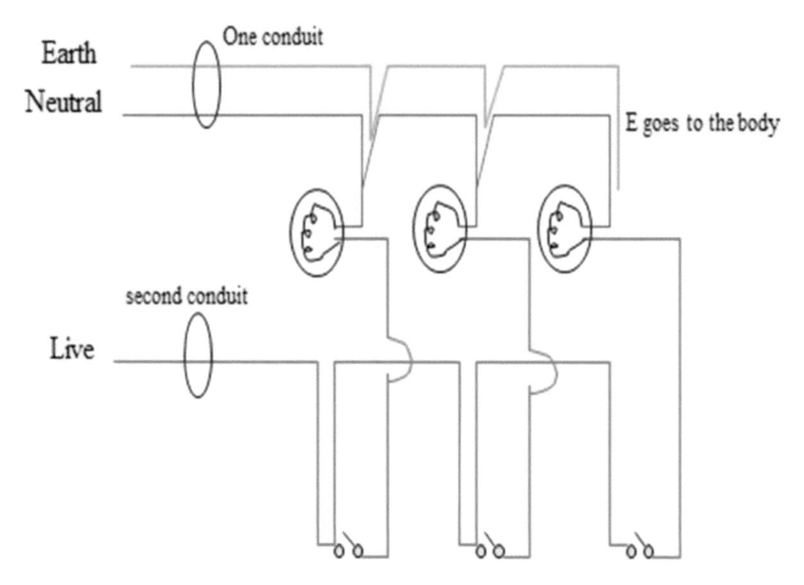


Fig. 81: Home wiring schematic

Table 7: Power dissipated in bulbs for different resistance of filament

| Incomming supply voltage = 240V | Voltage across each bulb | Resistance of bulb | Power disspated in bulb=$V^2/R$ | Number of times power dissipated in parallel bulb is higher |
|---|---|---|---|---|
| Series connection of two bulbs | 60 | 106 | 34 | |
| Parallel connection of two bulbs | 120 | 106 | 136 | 4 |
| Series connection of two bulbs | 240 | 160 | 360 | |
| Parallel connection of two bulbs | 120 | 160 | 90 | 4 |

Table 8: Experiment with 3 Φ voltage

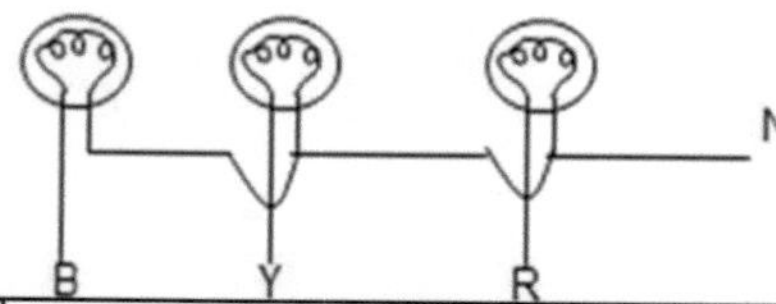


| Action | Voltage at terminals of bulb | | | |
|---|---|---|---|---|
| | B bulb | Y bulb | R bulb | Bulb parallel to R bulb |
| RYBN looped on one side of bulb | 241 | 241 | 237 | |
| N disconnected | 247 | 240 | 232 | |
| N disconnected and B bulb taken out | 360 | 216 | 200 | |
| N disconnected and B bulb taken out, R bulb connected in parallel to another bulb | 371 | 292 | 124 | 124 |
| YR = 417V | | | | |
| YB = 415V | | | | |
| BR = 416V | | | | |

Each bulb connected to one phase and N disconnected

Fig. 82: Experiment with 3Φ voltage

N disconnected from supply and one bulb taken out so only two phase supply. Thus this is a series circuit with two phase (R, Y) powering two bulbs but in parallel to one of these two bulbs is another bulb. **Amps determined by single bulb in series which will burn lower wattage equipment.**

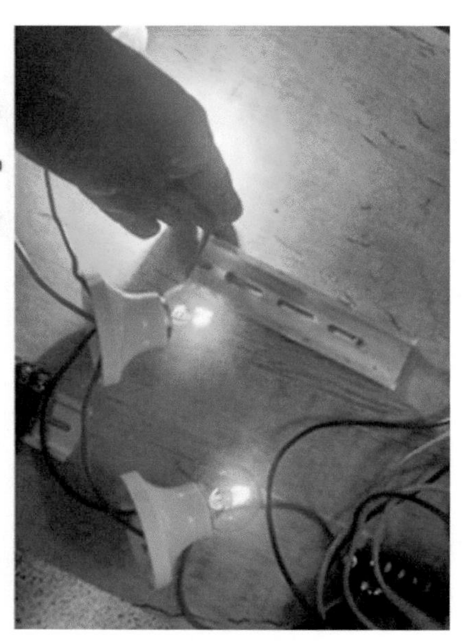

Fig. 83: Further experiments with 3Φ voltage

# Chapter 17

# Electric Cars

Electric cars will greatly increase electricity consumption over the next few years. Tesla Motors is spearheading electric car development while Google is turning electric cars into self-driving cars. Self-driving cars can travel even an inch from each other enabling existing roads to ferry many more cars that it can currently handle. Countries like India which has much less roads than she needs may not need to build all the extra roads. Accidents will be 90% less so hospitals and insurance business will be affected. Overall it is a disruptive technology which even Warren Buffett who is considered the most successful investor of the 20th century is concerned about because he has investment in the businesses which driverless cars will disrupt.

Specifications of Electric sports car made by Tesla Motors

- 0-60 (0-97km/h) in less than 3.07 seconds.
- 125 mph top speed (201 KM/h).
- Redline at 13,000 rpm.
- 621 miles (1,000 km)
- Full charge in 9 hours.

Zero emissions.

- Battery voltage: 375 volts.
- 3-phase, 4-pole electric motor, 248hp peak (185kW), redline 13,000 rpm, (having the name of Tesla they have to use Nikola Tesla's invention).

- Regenerative engine braking
- No camshafts.
- No engine block.
- No turbocharger.
- No supercharger.
- No lubrication system.
- No radiator.

In 2013 the largest electric car sales in U.S. was by Nissan followed by GM, Toyota and then Tesla. In 2018 Tesla is the top seller of electric cars. Tesla's cars are a revolution in incorporating in the cars, the very latest computer and cell phone technologies. Some are saying the name of the company honoring the greatest scientist of the world is a boon to the company. How can such a small startup like Tesla beat the giants like GE, Toyota or Volkswagen in their game? It must be that these old companies are dragging their feet into this new technology because they are too heavily invested in the old technology. Powerful as these giants are there is no stopping electric cars because government have already set targets for all cars to be electric. India, China, Indonesia, France and Britain have all set target dates for all cars in their countries to be electric.

The Tesla Model S has a motor running a gearbox and this gearbox is connected to the wheel. The gearbox has only two gears and it is fixed. These gears are not used to change the speed of the car as in ICE cars. The change of the apeed of the car is achieved by manipulating the sine wave of the AC going into the car motor. A long wavelength wave results in slower speed and a short wavelength results in a higher speed. A higher amplitude AC wave results in more power for the motor and low amplitude results in lower power. The two gears are meant to reduce the speed and equivalently increase the torque. Induction motors optimally run at a speed too high for cars so the two gears are used to reduce the speed to one more acceptable for humans. Prehaps one day as self driving cars become standard, even this gearbox will not be necessary and cars will travel at a standard 200 miles per hour at a inch from each other just as jet planes seem to travel at much higher speeds that that next to each other in air-shows.

An electric car has a torque that is quite flat over its whole range of speeds compared to only one particular speed at which an ICE car can achieve a maximum torque.

The most expensive portion of an electrical car is its battery but as with laptop and cellphone batteries, the energy density is going up and price is going down. There is also developments being made in induction motor technology such as using High Temperature Superconducting (HTS) coils of the motor. The most powerfull magnet in the world are made with HTS coils. The space in-between the rotor and the stator is also made a vacuum to achieve higher efficiency. From Nikola Tesla's first induction motor till today, the main improvement is the reducing the space in-between the rotor and stator because mechanical engineering (or accurate metal shaping) was not so precise in the 1800s. Engineers found that reducing the space in-between the rotor and the stator increased the power of the motor by a square. But if in addition, this space is made a vacuum, the energy density of the motor is increased. The only drawback is the need for a cooling system to cool the coils down to about 90°K (-183°C). But as HTS are getting operatable at higher-and-higher temperatures motors can be made multiple times smaller and therefore multiple times lighter. Of course the ultimate goal of superconductor scientists are to achieve a room-temperature superconductor material. When that happens the only 'engine' weight of the car will be the battery. But as we can see from our laptops or cell phone batteries are also getting smaller and therefore more power dense. One U.S. navy ship is currently being run with a HTS motor plus vacuum in-between rotor and stator. The ship's motor can carry 150 times the power of similar-sized copper-wire-motor and is less than half the size of conventional motors. It reduced the ship's weight by nearly 200 metric tons, making it much more fuel-efficient and freed up lots of space for additional warfighting equipment.

Battery developments in electric cars is not just battery chemistry but according to a top person in a high-end German battery manufacturer, the intelligent utilization and coolling of the thousand of battery cells placed at the bottom of the cars is much more important. That person stated that battery chemistry is known by all but the cooling system is now an industrial secret. For example, on one day of driving, a certain sets of battery cells are used and on another day another set of cells are used and so forth. In Tesla car,batteries are small and shaped like the ones in supermarket. A system of glycol cooling system wriggling through all these batteries is used to cool them more effectively than if a big battery is used. There are actually 7000 cells in a Tesla Model S. The glycol dissipates heat via a radiator fitted at the front of the car just as air-intake is currently fitted at that position in ICE cars.

Most people who are not mechanical engineers will not know that it was a great innovation to enable one internal combustion engine (ICE) to drive a car. This is because when humans moved from bullock carts to ICE cars there was a problem. An ICE cannot be joined to two wheels because then the two wheels will move at exactly the same speed and therefore cannot turn. When a bullock cart turns, the wheel at the outer diameter is turning faster than the inner diameter one. It was the invention of the differential that enabled one wheel to turn faster than the other. Today's ICE cars use a an open-differential instead of a limitted slip diffential used by advanced ICE cars. The limited-slip differential is used by advanced ICE cars to enable running over a puddle of water where the wheel which has traction with the road is turning but the one on the puddle is only turning a little. But the open-differential is utilized in electric cars because selective breaking of one of the front wheel can be done or even a complete stop of the induction motor can be done. A complete stop of an ICE is not possible.

Finally there is regerative braking. If a solenoid magnet is passed a wire at right angle, current is induced. If the wire is moving forward and the magnet is also moving but faster than the wire also current is induced in the wire. But if the instead, the wire is moving faster than the solenoid, current will be induced in the solenoid. Summarily when the solenoid is moving faster than the wire, current is induced in the wire and it builds its own field. If both are moving at the same speed neigher will experiecne a varying Field density but as the wire moves faster than the solenoid the solenoid experiences a varying magnetic Field density from the wire, while its own seem stationary to the wire. So when a human presses the brakes of a Tesla car, the RMF going into the stator coils are adjusted to be slower than the squirrel cage rotor speed, thereby generating current in the stator coils which goes to charge up the battery. In ICE cars, the brake just dissipates all this energy into the atmosphere as heat while in electric car it is retrieved.

A petrol station for an ICE car can cost many millions but to make a charging station for an electric car is just pulling wires and placing a transformer in the charging station. The total cost could be just a thousand or two. The transformer is placed here so as to enabling faster charging. For the Tesla Model S, it takes 20 minutes to charge it to half full.

One of greatest advantage of an electric car is that eventually all car parks will have a plug point. Thie is not strange or never-been-done-before thing. In Scandinavian countries and where this author studied in South Dakota, USA, all car parks has plug points even in the 1980s. Those plug points were not for electric cars but to keep cars warm so that it does not freeze in the upto -90°F climates common in South Dakota during Christmas. So, with most cars plugged in from 8am to 6pm, this is a huge battery bank for power companies. In the state this autoor lives, the spinning reserve or running of extra generators to cater to a big generator tripping is 30%. Meaning 30% extra power is generated that is not utilized. If the only source of power is gas turbines or coal plants that is 30% extra pollution. With all electric cars plugged in most of the day, that whole 30% fuel can be saved and car batteries will be the backup. Batteries are the fastest backup there is. In banks only battery banks are fast enough to prevent their critial computer systems from going down.

Another great advantage is the number of parts in an electric car. There are reports of the number of parts are about 150 while ICE cars require upto 10,000 parts. Even a human can keep up with 150 parts in their minds and need not have a computer to figure out parts level. A computer will definitely be needed to keep track of 10,000 parts. So eventually the production cost of electric cars will be so low, ICE cars will not be able to keep up.

# Chapter 18

# Superconductivity

Superconductivity is zero electrical resistance and zero magnetic fields occurring when certain materials are cooled below a certain temperature. This property was discovered in 1911. The ions (atoms with less than normal electrons since conductors love to give off electrons) within the conductors just stop vibrating so electrons can glide over them without any resistance like a MAGLEV (Magnetically Levitated) train travelling over a magnetic field. For copper or silver there is some electrical resistance even near 0 K. But in superconductors the resistance drops abruptly to zero below the critical temperature.

Superconductive materials like mercury needed to be cooled to near 0K to achieve superconductivity. Then in 1986 a material was created to have superconductivity at 90K (-183°C). So, such materials are called high-temperature superconductors (HTS). The highest temperature superconductor to date is mercury barium calcium copper oxide ($HgBa_2Ca_2Cu_3O_8$) at around 138 K (−135 °C). Because liquid nitrogen boils at 77K (-196°C) many applications are now possible with HTS, including:

- MRI machines for hospitals (the most common)
- NMR machines for mass spectroscopy
- Mass spectrometers
- Beam-steering magnets used in particle accelerators
- Magnetic separation; to separate weakly magnetic particles from less or non-magnetic particles (in pigment industries)
- Motors

- Magnetic levitation devices (maglev trains)
- Energy storage

Future uses of superconductors includes:

- Electric power transmission
- Transformers
- Fault current limiters
- Superconducting magnetic refrigeration.

Superconductivity is sensitive to moving magnetic fields so applications that use AC like transformers will be more difficult to build that those that works on DC. Superconductors can maintain a current with no voltage and this fact is already utilized MRI machines where the magnetism is maintained even though the power is switched off. Superconducting magnets are some of the most powerful electromagnets known. Experiments have shown that currents in superconductors can continue flowing for up to 100,000 years.

# Chapter 19

## Anode and Cathode

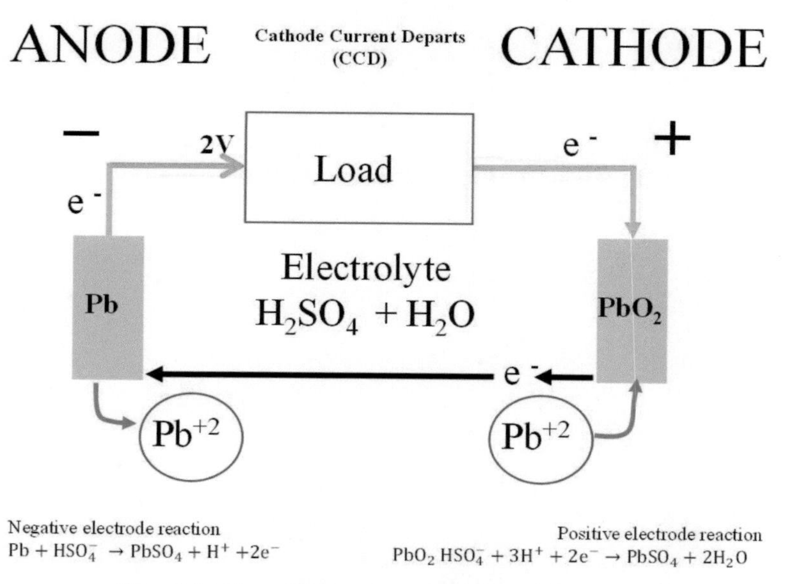


Fig. 84: Pb acid battery Anode and Cathode

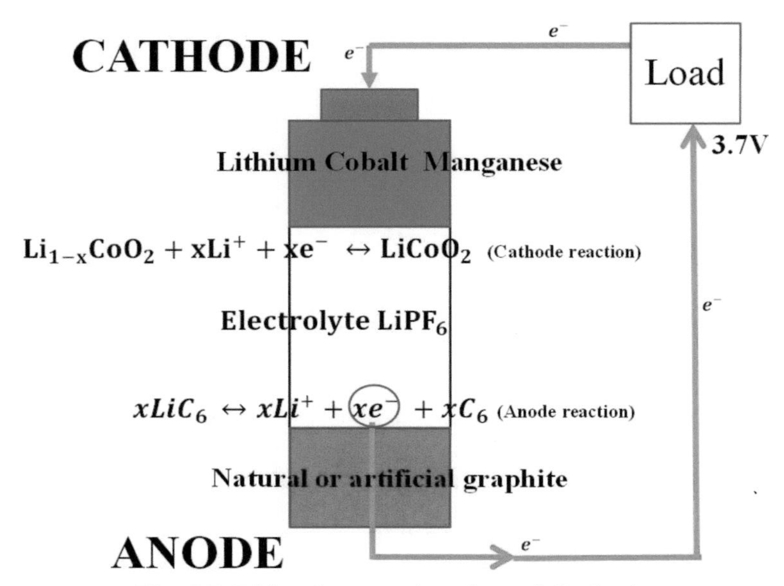


Fig. 85: Li ion battery Anode and Cathode

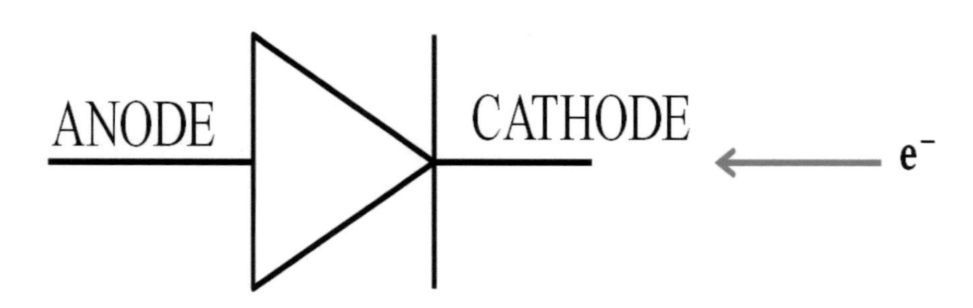


Fig. 86: Diode Anode and Cathode

Fig. 84, Fig. 85 and Fig. 86 depicts the definition of anode and cathode which are words created by Faraday. In a Li ion battery, Li ions move from the anode to the cathode. A cathode is an electrode through which electric current flows out of a polarized electrical device. A mnemonic used to remember this is **CCD (Cathode Current Departs).** Note current flow is in opposite direction of electron flow.

An anode is an electrode through which electric current flows into a polarized electrical device. The mnemonic used to remember this is **ACID (Anode Current Into Device)**. In a galvanic cell (standard battery from a shop), the system has stored energy and if the circuit is completed, the system would like to discharge, therefore electrons (negative) will flow from the negative electrode (which has the "minus" sign) via the wire to the to the positive electrode (with the "plus" sign).

# Chapter 20

# Rectifiers

The original grid of Edison was DC so when it got switched to AC, rectifiers (which convert AC to DC) had to be developed. The simplest rectifier is the transformer bridge diode, capacitor system shown in Fig. 86. It works according to Fig. 87.2. A diode will cut off the bottom of an AC wave as shown in Fig. 87.2 top left, second graph. A bridge diode circuit provides a full wave rectified waveform as shown in Fig. 87.2 bottom first graph. But if a capacitor is connected at the output, the graph will look like the bottom second graph. As the wave is going down, it is stored in the capacitor which releases the energy half way down the slope, resulting in a much flatter as shown in the bottom most graph.

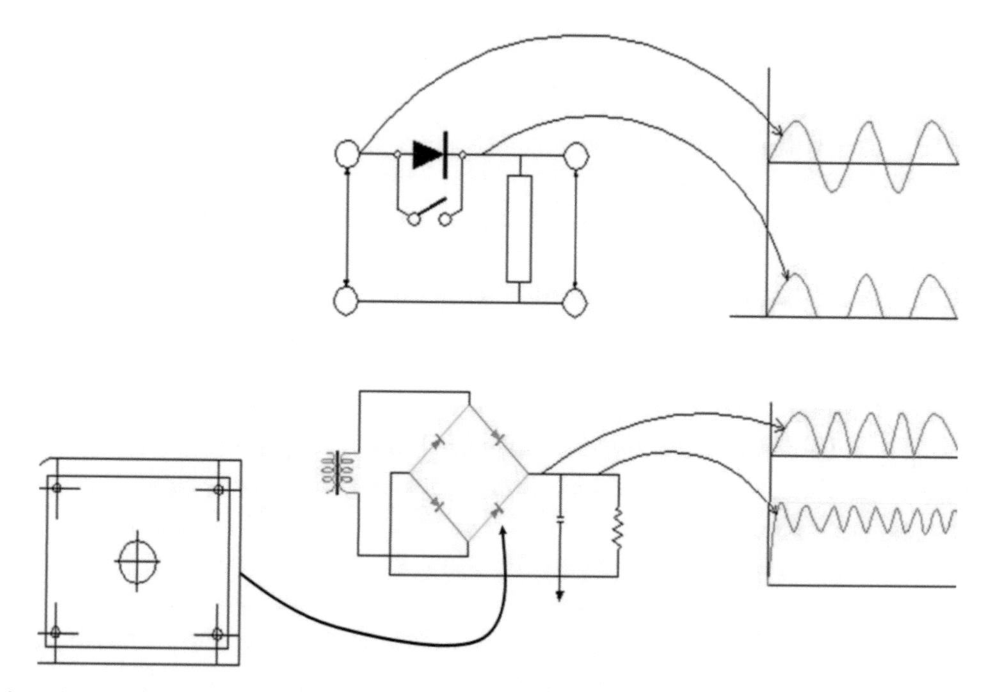

Fig. 87.2: Full bridge rectifier circuit, box on left is how rectifier currently looks like

## 20.1 Switch Mode Power Supply (SMPS)

The rectifier of Fig. 87.2 had a DC output that was alright for devices up to the 1980s and 1990s but is too wavy for the latest cell phones, radios, TVs and so forth. So, the Switch Mode Power Supply (SMPS) is currently used almost universally as the main power supply for all electronic appliances.

The physical SMPS is shown in Fig. 88 and Fig. 89 is the schematic and Fig. 90 is the waveform sequence. It works on the principle of first rectifying the incoming AC which is 120V (240V British) and then chopping the resulting DC it into a high frequency up to 100kHz for the latest cell phones but up to 1000kHz has been created in labs. This high frequency 120V AC (240V AC British) at 100kHz is then sent to the transformer and stepped downed to 5V AC at 100kHz for a typical cell phone charger. Then comes the rectifier and capacitor which are used to rectify or convert the AC to a very flat 5V DC. The capacitor has to be chosen to store energy for only half the time an AC square wave takes to go down to zero and then release its charge at that time, making the wave start going up after it has gone down half way. At 100kHz, this is a very short time, resulting in a much flatter wave than the old system of Fig. 87.2. So, the higher the frequency of the 5VAC, the flatter the 5V DC will be. Note 5V AC will be transformed to 5V DC but there is some voltage drop across the four diodes of a bridge rectifier. Most diodes have a voltage drop of about 0.7V per diode. How is it that 5VAC is transformed to about 5VDC? The story starts from the early days of electricity. Edison created the DC electric supply grid and its main function was initially to heat water in the basement of tall buildings. This heated water was then piped to each room in the building where they spin in a spiral replacing old fireplace; which tended in those days to catch fire and burn buildings. So, when JP Morgan decided to move to AC to heat this water, the technicians were not willing to learn a new number of the AC system. How they determined the equivalence of AC to DC was by simply heating two same containers of water. One was heated with a heater supplied with DC and the other heated by a heater supplied with the new AC. If it took 10 minutes to boil water with DC 120V, they determined that it will take 170V AC peak voltage to boil the same amount of water with AC in 10 minutes. But Tesla said its alright, they can still use the same number used in DC by just multiplying the AC peak voltage of 170V with 0.707 to give the AC RMS (Root Mean Square) voltage of 120V. So, DC 120V has the same heating power as AC 120V RMS. 0.707 can be remembered because there was a Boeing plane called Boeing 707. Meters can only measure the peak voltage but today the measurement is multiplied by 0.707 on the scales of the analog meter to read the RMS value. On digital multimeter, a simple multiplication of peak voltage by 0.707 is done to give the RMS voltage.

As can be seen from Fig. 87.2 of a typical non-SMPS rectifier; a single diode on the top left schematic just cuts off the bottom half of the AC wave. The resultant wave has a time period of half cycle where there is no voltage. This is enabled by the property of a diode to allow current to flow only in one direction.

On the bottom Fig. 87.2 is a full wave rectifier where four diodes are used instead of one. This enables voltage to be present over the full time period. This bumpy wave is still DC because it does not go below the 0V line. Note, even a full sine wave above 0V is still DC but a noisy one. To change this bumpy DC into a flatter DC, a capacitor is attached parallel to the output leads. This capacitor is chosen to absorb current over ¼ cycle and release it over the next ¼ cycle. Thus, instead of the voltage going down after the ¼ cycle it stays up and goes down only slightly. The resulting waveform is shown on the bottom right graph.

The reason the SMPS 5V DC is smooth can be observed if one looks at the bottom right of Fig. 87.2 top graph. The normal rectifier of a power company 60 Hz (50 Hz British) AC wave has the DC going down before being pushed up by the stored energy in the capacitor providing the waveform of the Fig. 87.2 bottom-most graph. But if the frequency is 100,000 Hz as in the latest cell phone charger SMPS, the waveform can only go down a tiny amount before the capacitor release of energy pushes it back up since each wave is packed very close to each other with respect to time. The result is a much flatter DC waveform. In labs 1MHz switching of SMPS has been achieved. Of course, the flattest DC is from a battery.

Another big advantage of SMPS is the fact that for any transformer the size goes down with increase in frequency. Thus, SMPS of even high amp (note wire size only depends on amps) equipment is quite small. This enables saving of copper whose price have gone up tremendously over time. Thus, the latest cell phone chargers are small (small transformer) and outputs a very flat DC. As an indication of the size difference caused by using high frequency waveforms; a standard CPU power supply shown in Fig. 88, the label on the outside shows 450 watt and the 3.3V, 5V and 12V transformers (green top) are rated 28A, 40A and 20A respectively. Their sizes are half inch cube and smaller. Comparatively while this author was building a project at university in USA in the 1990s, a 6A, 12V output normal transformer was used, measuring about 7" cube. The weight of that single transformer was much heavier than the whole computer supply of Fig. 88. So, if this computer used the same old technology transformer, the transformer itself should be 7" X 20A/6A ≈ 23" cube! ≈ a two feet cube box.

Fig. 88: Switch Mode Power Supply of computer

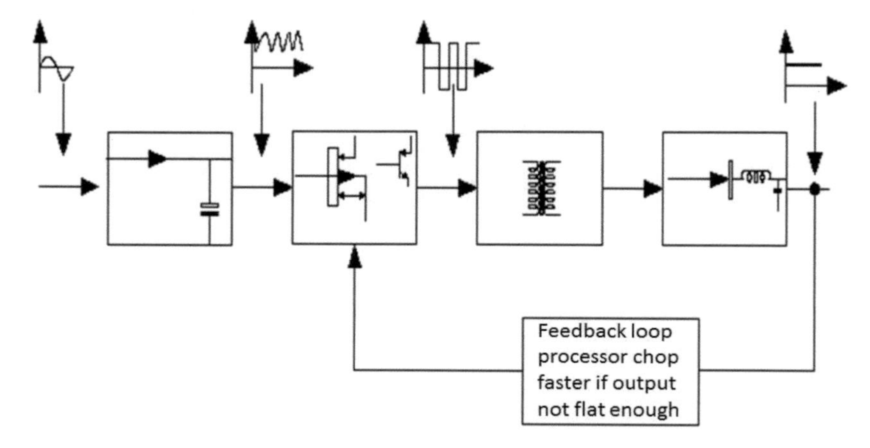


Fig. 89: Switch Mode Power Supply block diagram

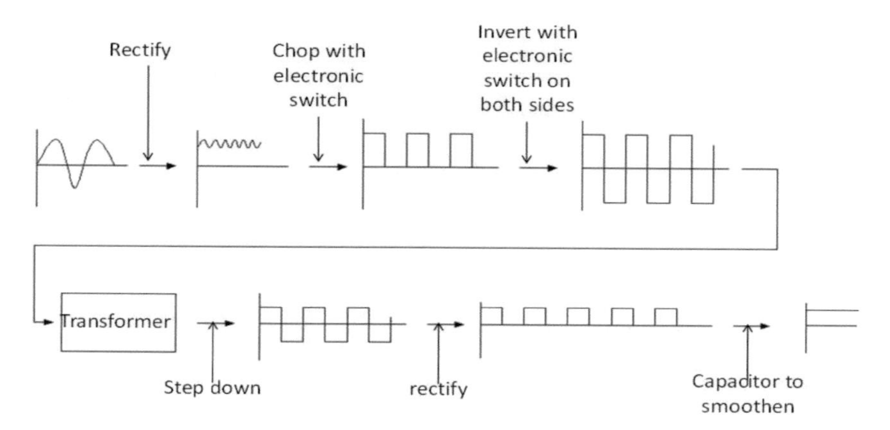


Fig. 90: Switch Mode Power Supply waveform

## 20.2 Detecting rectifier terminals

Detecting rectifier terminals is a very important. In many industries, specifically aluminum smelting, huge rectifiers are used. They are also used to produce the DC to drive the rotors of generators. Mistakes in detecting their terminals can be very expensive. The procedure for detecting transformer and rectifier terminals are as below:

Detecting rectifier terminals: Fix Red multimeter probe on one terminal, Black multimeter probe touching the other three terminals one by one:

Ohmmeter always deflect = Red probe on + terminal

Ohmmeter always do not deflect= Red probe on - terminal

Ohmmeter sometimes deflect = Red probe on AC terminal

Transformer - two terminals have high ohm = join incoming L,N to these terminals

Transformer - two terminals have low ohm = join outgoing to rectifier AC to these terminals

If transformer has 5 terminals, the fifth one is joined to the temperature sensor. It has no continuity to any other transformer terminals. Note the high voltage side of the transformer always has the thinner wire size coils and the low voltage side of the transformer has the thicker wire size coils. There was an expensive mistake where a $10 million transformer was destroyed because the high voltage was connected to the larger hole size and the low voltage was connected to the lower hole size. It should be opposite, large wire for low voltage and thinner wire for high voltage. Else just remember that wire size is always dependent only on amps and not on voltage or power.

The coil wire size of a transformer is designed just like any other wire design. First sum up all the loads in the installation in watt, then use $P=VI\cos\theta$ (for single $\Phi$) or $P=\sqrt{3}\cos\theta$ (for three $\Phi$) to calculate the current. From this current the wire size can be determined from the current level versus wire size table. This table is used to determine the correct wire size which will not burn at that current level.

Special cases include long distance between source and load ($V_D$ calculation need to be done), very hot environment ($C_a$ need to be calculated), Compact space for running wires ($C_g$ need to be calculated), abnormal insulation ($C_i$ need to be calculated). All these calculations were detailed earlier.

# Chapter 21

# Power Generation

Electric power generation is one of the main polluters of earth, emitting 32% of $CO_2$ pollution of earth. The whole transportation industry emits 25% of $CO_2$. So, more research and technological development need to be invested in attacking the pollution of the electric power industry. The current world electric power sources are shown in Table 9 and Fig. 91.

As a side note to the electrical industry, a mostly neglected fact is that even though the farming industry, as shown in Fig. 92 produces only 10% of $CO_2$ production worldwide, the animal industry produces between 14.5% (U.N. figure) to 51% (Worldwatch Institute) of global warming pollution because there are many other global warming gasses other than $CO_2$. But no one doubts that it is more than the whole transportation industry pollution or more than the combined emissions of all the cars, trucks, trains, planes and ships. In the U.S., the animal rearing industry emits 9% of global $CO_2$ (only 1% by non-animal farming), 35-40% of methane, 65% of nitrous oxide and 80% of ammonia. Added to this, an estimate 33% of fossil fuels consumed in the U.S. is utilized for rearing animals, meaning 33% of the pollution emitted by the transportation industry must be added to the pollution caused by the animal rearing industry.

Coal (utilizing steam turbines (ST)) is still the main source of electric power producing 40% of it. 37 % of electricity is produced by gas (using Gas Turbines (GT)) and 9% nuclear (using ST) and 4% is produced by oil (using internal combustion engine (ICE)) another 4% with solar and 2% each from wind and others like geothermal and other plants. ST which is used by coal power plant and nuclear power plants generate over 50% of electricity. The steam turbine was first invented in UK. A simple description of how it works is by filling an oil-drum with about ¼ full with water. A flame is placed below this oil-drum which will turn a portion of the water in the drum into steam. Then a little cool water is sprayed into the steam, condensing some of it to water which will create a vacuum within the drum which will collapse the drum. A strong over 2mm thick oil-drum which a human cannot even bend a little will be totally collapsed. So, if the drum is made even stronger, it can pull a piston down, thereby doing work. Such a system is perfected to make the original steam engine.

Many top thinkers in the electric field believe solar is the future because it is so simple. A ST in a coal plant, a GT in a gas plant or an ICE in a oil plant are quite complicated. ST for example have fins that must be positioned extremely precisely (1/40000 inch = 0.00025" = 0.00635mm) near the housing structure for them to function efficiently; a slight movement out of the fins can reduce the efficiency of the GT or GT drastically. Bearing have to be continuously monitored to ensure perfect functioning. A mechanical damage or a small splinter going into the tight spaces in-between GT fins and housing can cause havoc or even a powerful explosion.

Comparatively a solar panel takes out all that complication and seems to have done what microprocessors have done to many previously mechanical machines. Of course, there is precision and very high-end science in the manufacturing of solar panels but for the operator or maintenance personnel it is just plug-and-play. Solar panels use the very latest pure silicon used by the highest end computer chip manufacturing to achieve highest efficiency of converting sunlight into electricity.

Table 9: World electric power by source (TWh/year)

| | Coal | Gas | Hydro | Nuclear | Oil | Others | Wind | Solar | Total |
|---|---|---|---|---|---|---|---|---|---|
| Actual GW | 8390 | 4744 | 3619 | 2344 | 879 | 541 | 501 | 105 | 21123 |
| % of world usage | 39.72 | 22.46 | 17.13 | 11.10 | 4.16 | 2.56 | 2.37 | 0.50 | |

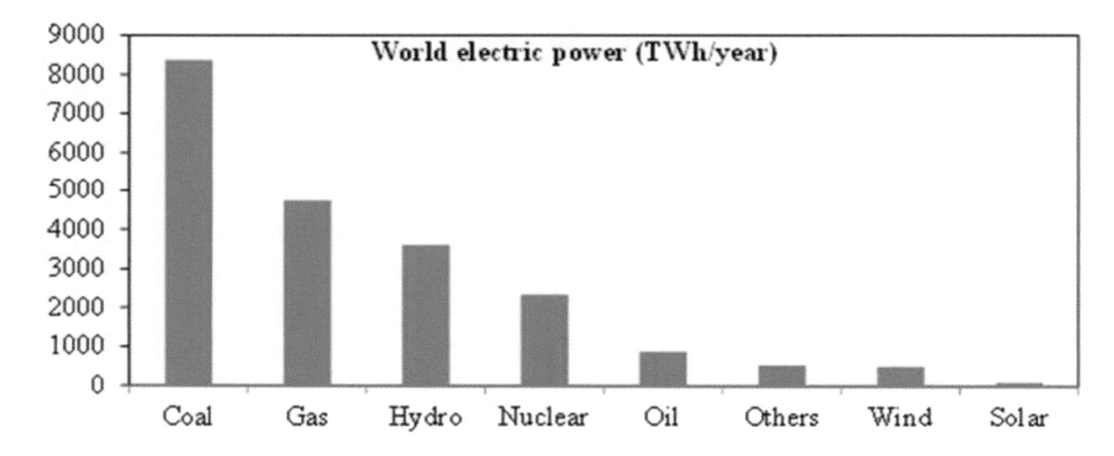


Fig. 91: Electric Power generation sources

Of the total energy produced, 66% is generated from hydrocarbon sources. Specifically, 40% from coal, 22% from gas and 4% from oil another 4%. Solar produces 4% and wind and geothermal produces 2%. There is much excitement in the solar power arena because they have reached grid parity, meaning the cost per kW is equal to electricity from coal in some high electricity price regions like California and Hawaii. But solar power is not for all countries. The most successful solar projects are located in deserts or low cloud cover regions. In tropical or equatorial countries where intermittent clouds comes and goes, it is not an optimum solution. The clouds in equatorial regions look thin but taking a plane, one can observe that it can be as thick as a tall mountain. That will heavily reduce sunlight reaching solar panels.

The problem of solar power utilization in cloudy countries is depicted by the experience of Hawaii in 2015. Dependence on solar power damaged grid equipment because a **cloud cover can take out 70 to 80 percent of solar power output in less than a minute.** Thus, the other power stations and switchgears in Hawaii were damaged as the load drew all their demand from the remaining diesel engines and gas turbines. Other than hydroelectricity, no other power stations can load the grid or backup this loss that fast. One of the generators of the Bakun dam in Sarawak can load 300MW to the grid in 20 seconds. Therefore, it is optimum for a tropical and cloudy country to develop both hydro and solar power simultaneously. This author's previous job at a power company was to call the various power stations to start up extra generators or shut them down. When a GT was called to start-up, it would take 45 minutes before the electrical power got onto the grid. For an ICE it would take more than 60 minutes, a coal power station can take eight hours but a hydro power station can load the grid with power in 20 to 30 seconds. Therefore, only hydro power station can back-up a sudden cloud cover over a huge solar farm.

Fig. 92 depicts the $CO_2$ emission by various sources in 2013. The electric power industry produces 32% of the $CO_2$ emission to the atmosphere. This is a total 6,526 million metric tons of $CO_2$ emitted to the atmosphere every year and it is increasing every year. Fig. 93 depicts the feed stock for generating electric power currently. Table 10 depicts the top electric power generation countries.

Solar power has to be harnessed as much as possible in countries with little cloud cover located mostly in temperate or desert regions. A pure desert which has the least cloud cover has a problem with solar power currently because heat affects solar panel's efficiency. Hopefully innovations can solve this problem. The biggest problem with solar power is storage because there is no energy production at night. But Elon Musk of Tesla Motors recently claimed he has solved the critical battery problem with what he named, 'Powerwall".

## 21.1 Justification for Hydro power

Wind power need to be installed in heavy wind areas like the Mid-West of USA and off the coast of Norway and North Sea. And solar power plants are optimally located in cloudless but cool regions of the earth. For the equatorial regions where rainfall is typically 560% of the average level in the Northern hemisphere, hydropower should be the optimum. But hydropower requires another factor which is high mountains to enable good gravity drop of the water. For example, this author lives in Sarawak, Malaysia which has optimum conditions for hydro power namely:

- Sarawak is the largest state of Malaysia (at 124,450 km$^2$)
- Sarawak has a population of only 2.5 million so large areas can be dammed up.

- Sarawak is located on the equator and therefore has a rainfall of about 4000mm (157 inches) per year compared to a USA average of 715mm (28 inches), which is 560% higher rainfall than USA.
- Sarawak has a topography of high mountain ranges at the border with Indonesia.

Brazil is also a rainforest but lacks mountains so the Itaipu dam needed to be built at the border regions with Paraguay. Thus, the Itaipu dam's power is shared with Paraguay.

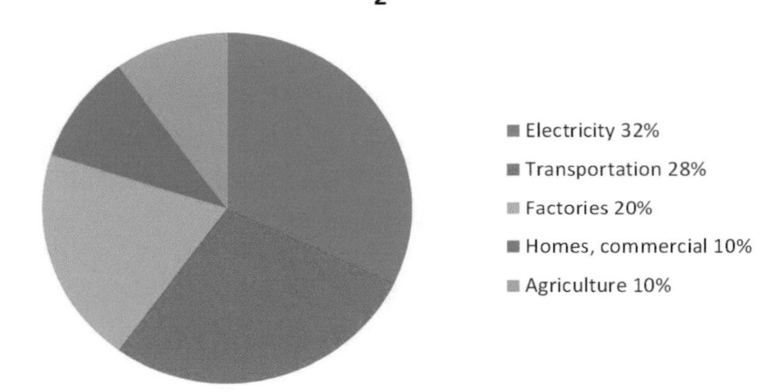


Fig. 92: $CO_2$ emission by various sectors

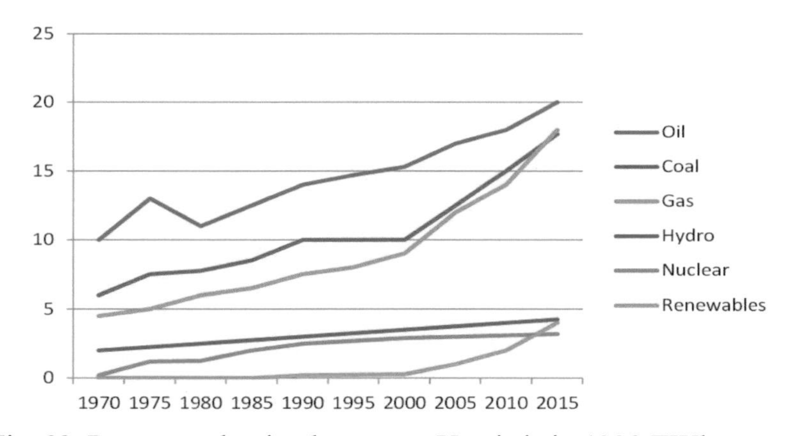


Fig. 93: Power production by source Y-axis is in 1000 TWh per year

Table 10: Electric power production by country

| Rank | Country | Electricity consumption (MWh/year) |
|---|---|---|
| 1 | China | 6,483,800,000 |
| 2 | United States | 4,686,400,000 |
| 3 | India | 1,111,722,000 |
| 4 | Russia | 1,016,500,000 |
| 5 | Japan | 859,700,000 |
| 6 | Germany | 582,500,000 |
| 7 | Canada | 499,900,000 |
| 8 | France | 462,900,000 |
| 9 | Brazil | 455,800,000 |
| 10 | South Korea | 455,100,000 |
| 11 | U.K. | 323,300,000 |

## 21.2 Wind power

For wind turbines, GE's advanced technology allows the elimination of gearboxes. Direct drive or gearless turbines reduce the number of moving parts in a unit and increase reliability, helping to minimize costly open-sea maintenance.

However, Warren Buffer said industrialist will get a lot of tax credit if they build a lot of wind farms which is the only reason he invests in them. And Bill Gates explained in an interview with the Financial Times why current renewables are dead-end technologies. He said they are unreliable because battery storage is inadequate and wind and solar output depends on the weather. Bill Gates stated that the cost of reducing $CO_2$ emission using today's technology in wind and solar are 'astronomical'. Wind power turbines are especially notorious in killing birds. The figure can reach as high as 80 birds per turbine per year. Therefore, the total number of birds killed per year by wind turbines is estimated to be as high as 39 million. It should be noted that wind power harnessing totally does not work in the Equatorial regions of the earth. Wind is high at the Northern and Southern Hemisphere and very low at the Equator. In the old days British sailors called the equatorial regions "doldrums" because the wind speed is so low that they have problem sailing and could be stuck in these regions for long periods. In Equatorial Sarawak, a wind turbine was installed at the tail end of the Borneo island; this spot is surrounded by the south China Sea but it did not turn. Then another turbine was installed on top of a mountainous portion of Borneo which should have lots of wind but this also did not turn.

## 21.3 Wind power versus the rest

Fossil-fuel plants are cheaper now, but if the UN charter of paying for $CO_2$ emissions are included, wind farms are cheaper. Also, the price of oil can go up drastically in the near future. An equivalent power output deep-water wind turbines and nuclear power station cost about the same. This is also equivalence in price of coal power station with capture and sequestration and wind turbines. Deep water wind farms are only slightly more expensive than onshore wind farms. Deepwater wind turbines are normally 47 to 70 km off-shore, are normally four legged but there is one unit in the world that floats on the water. There is always more wind the greater the distance from the shore. Some researchers have also envisaged flying balloons with wind turbines mounted on them. The higher off the ground, the higher the wind speed.

## 21.4 Fukushima Nuclear Disaster

Professor Emeritus Kiyoshi Kurokawa from Tokyo University and his team which investigated the Fukushima disaster made a statement that Fukushima cannot be regarded as a natural disaster and it was a profoundly man-made disaster that could and should have been foreseen and prevented. He also said its effects could have been prevented by a more effective human response to the disaster. He said governments, regulatory authorities and Tokyo Electric Power [TEPCO] lacked the sense of responsibility for the lives of the people and society. He effectively concluded a bit emotionally that the TEPCO effectively betrayed the nation.

## 21.5 Solar energy

Photovoltaic (PV) technology has sprung up recently all over the world greatly aided by cheap solar panels made in China. The idea of using crystals or non-metals in the electrical industry began with Jagadis Chandra

Bose in 1895. Bose used it to build a radio receiver. The silicon industry has now grown into the computer/cell phone industry, the fiber optics industry and the PV solar industry.

Elon Musk of Tesla motors stated that it takes only a small square portion of Northern Texas to power the whole U.S. with solar power. And the batteries required are a tiny portion of this small square. He is building a Gigafactory in Nevada to enable this. Of course, he is not proposing that all the solar panels be located in Northern Texas though if the U.S. government decides on this route, Northern Texas has an opportune low cloud cover plus cool temperature to enable this; cool temperatures are good for PV efficiency as well as battery life. Elon Musk idea is that the solar panels should be widely sold in individuals' homes together with efficient computer controlled, batteries he calls the, 'Powerwall'. Elon Mush proposed solar panels to be on top of homes and office so it will not take away crop land. Batteries are the biggest reason for the failure of solar power projects. Even deep cycle lead-acid (Pb acid) batteries tend to last a maximum of three years. While nickel cadmium (NiCd) batteries can last for 20 years but tend to cost about five times in initial investment. But considering the utility of 20 years, it is worth purchasing nickel cadmium batteries, what is the use of a lead-acid system that last only slightly above two years? The high price of nickel cadmium is the reason most contractors choose lead-acid batteries for all solar projects so they can cut cost and collect money for their project fast. But the project is guaranteed to fail in two to four years. Lead-acid batteries have been used for a long time in cars where after continual research, they have the lowest internal resistance (50mΩ) but they do not last long. NiCd batteries have a 20-year life-span and has an internal resistance (70mΩ) almost equal to lead-acid batteries. Other major battery chemistry are Li ion (320mΩ) and NiMH (778mΩ), both of which have much higher internal resistance. The internal resistance is the resistance which causes a battery to discharge even when placed on a shelve because the internal resistance is the resistance in the chemical makeup of the battery. Since NiCd batteries has almost as low internal resistance as lead-acid ones plus it has a long life-span, and can withstand deep discharge, it should be the best. The only problem is that Cd is banned in Europe for health reasons.

Internal resistance is what causes the terminal voltage of 12V car batteries to be about 14V when the car (or alternator) is running and below 11V, if it is disconnected from the car and a load is connected to its terminals. Car batteries push out up to 70A during start-up. But the battery is only used for start-up, after that, the alternator energizes the required electronics and lamps of the car. But if more than 20% of its charge is used, it starts to die permanently. Which is alright for car batteries where batteries are only used for cranking and after that the alternator takes over. Lead acid batteries have another problem in that it cannot stand a sudden discharge so it is not suitable to be used as a battery-backup in buildings where a power company outage will require a sudden discharge and more than 20% usage of its stored energy.

NiCd batteries on the other hand do not have all these problems. But they are about five times more expensive. The internal resistance of NiCd is almost equal to that of Pb Acid batteries which is the lowest for common battery chemistries. The most common battery chemistries are lead-acid (voltage per cell = 2V), NiCd (voltage per cell=1.2V), Li ion (lithium ion; voltage per cell=3.7V) and NiMH (Ni Metal Hydrite; voltage per cell=1.2V). But for a solar project to work NiCd is the best solution. Another problem is that batteries generally survive well only at 25°C and below which means in equatorial and tropical regions or during summers in the Northern and Southern regions air-conditioning for the batteries is a must.

Previously solar panels were made of a series of 12 or 24V arrays. Today 48V system is the norm which results is higher efficiency tapping of solar power due to lower line losses. Solar energy has become feasible option for many poorer countries because manufacturing in China has driven down solar panel cost world-wide by about 62.5% within three years (2012 to 2015). That is, the price has gone from $0.8 per watt to $0.3 per wat.

Another type of solar power is Solar Thermal plants. These power stations are opportune for very hot deserts. The high temperatures can only be better for these types of plants while the high temperature drastically reduce PV efficiency. In Solar Thermal Plants cheap mirrors (not solar panels) direct sunlight to a central tower to hear up salt to a liquid at 500°C. This hot molten salt in stored in tanks and can boil water to power a steam turbine even at night. This is a huge advantage in that we can finally have non-intermittent solar power. 69 locations world-wide have built such plants with power generating capacity of 4.3 GW. Some boil salt but most just boil water to run steam turbines so they do not work at night.

## 21.6 Tidal Power

Since 2008, a dual-rotor tidal turbine has been feeding up to 1.2MW of power to the electrical grid of Ireland. A 1320MW barrage is being built in South Korea with an expected completion date of 2017. Another 320 MW tidal power plant is being planned in Swansea, UK. As of today, a total of 16 tidal power projects are in various stages of being utilized worldwide.

## 21.7 Fuel Cell

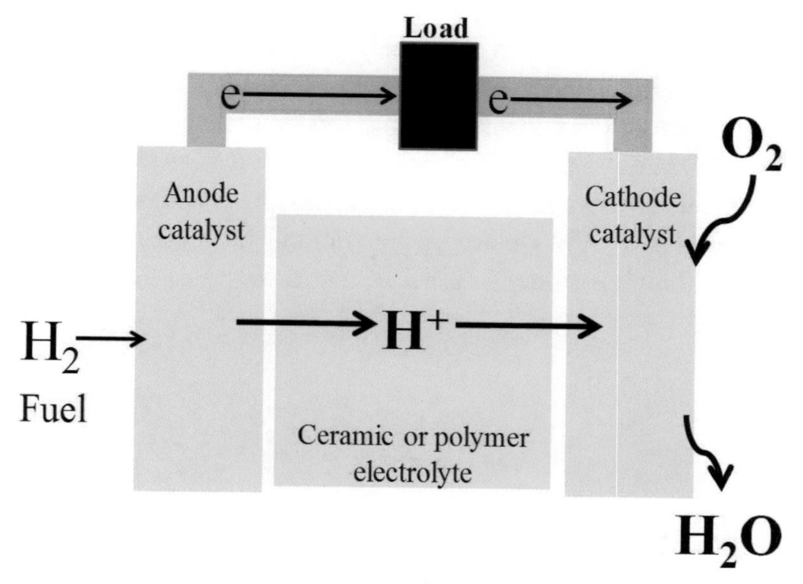


Fig. 94: Fuel cell schematic

At one time fuel cell technology depicted in Fig. 94 was mooted to be the future but it still uses fossil fuels. Basically, hydrogen is extracted from natural gas. Hydrogen is the simplest element in the universe with one proton and one electron. At the anode, the proton moves through the electrolyte and the electron powers a load like cell phone. At the cathode, the proton recombines with the electron and oxygen from air to output water. The big problem with this technology is that it still uses fossil fuels plus the energy to take out hydrogen from hydrocarbon is more than the output power that emanates from the fuel cell. It is currently well utilized to power servers in Silicon Valley by the likes of Google, Apple, eBay and many others.

## 21.8 Hydrogen

There are some people who are keen on the hydrogen economy. Two main methods to derive hydrogen are steam reformation of hydrocarbons and water electrolysis, the former being more popularly used. Hydrogen combustion is much more efficient than gasoline but unlike gasoline, hydrogen do not exist as a liquid at room temperature and at one atmosphere. That one problem brings about a whole slew of expenses. Taking out hydrogen from hydrocarbons or water actually utilizes more energy than the mechanical energy it generates upon combustion. Then there is the transportation cost in heavy gauge metal containers on trucks and at the pumping station due to the high pressure; thereby increasing the cost significantly. This container as well as the one on the truck must be maintained with zero leakage. If there is a leakage on a pump station a Hindenburg size explosion can occur. In hydrogen driven cars also, there should not even be a ¼ mm hole. The hydrogen fuel tank is not the only portion of the car that must be totally sealed, there are pipes sending fuel to the accessories and the engine; all must be totally leak proof. A small leak in any portion of the system will kill passengers in an explosion. Customers will be very worried about owning a car like this. So, overall it is not a practical solution for transportation. For electricity generation, since it takes more energy to take hydrogen out of hydrocarbons or water than it provides, it is not economical.

# Chapter 22

# The Air Conditioning system

Air conditioner (air con) is a heat pump whose schematic is shown in Fig. 95. It is a simple but ingenious invention which is very widely used worldwide. The inventor is Willis Carrier who actually got a degree in engineering in 1901. Basically, following Fig. 95, a coolant gets compressed (in the Compressor), liquified (in the Condenser), depressured (in the Expansion valve) and then evaporated (in the Evaporator). These are the four main subsections of an air con system. A liquefied natural gas (LNG) plant is huge and can reach many miles long but it is basically a huge air con system. Here natural gas (a combination of ethane, propane, butane, and pentane) from off-shore is sent to the plant in large pipes. In the plant, coolant gas (methane) flows in a smaller pipe circulating around the gas from off-shore to cool it to -161°C. The product of Liquid Natural Gas (LNG) is poured into specialized ships to be exported. The coolant gas of methane needs 316 atm (atmospheric pressures) to remain liquid and its boiling point is -161.5°C. Comparatively the coolant gas in a home air con or car air con boils at between -30°C to -50°C. Heat pump is the engineering term for the air con cycle (or refrigeration cycle). Basically, heat pumps and the opposite of it, namely heat engines (like combustion engines) are what drives human civilization today.

The main components of a refrigeration cycle are the evaporator, the compressor, the condenser and the expansion valve. The main formula for the refrigeration cycle is as below. This law is called the ideal gas law:

$$PV = nRT \quad (63)$$

Where:

P is the absolute pressure (Unit used: atmospheres, atm)

V is the volume (Unit used: Liter, L)

n is the amount of substance (loosely number of moles of gas)

R is the gas constant, which is the same for all gases

(Unit and number used: 0.0821 L·atm·mol-1·K-1)
T is the temperature on the absolute temperature scale (Unit used: Kelvin).

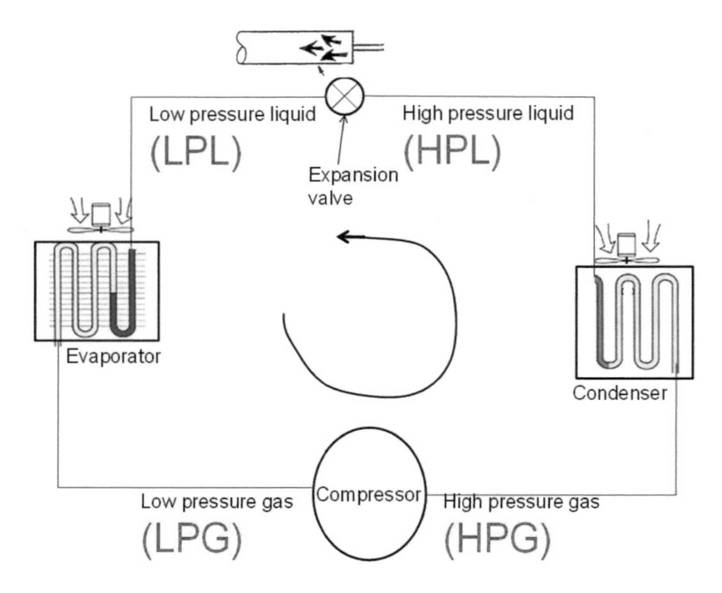


Fig. 95: The air-con cycle

Many students ask how to remember all these components, it is simple, just remember two things. The first thing to remember is that the compressor can only compress gas, it will be very hard to compress liquid; for example, fill a balloon with water and try compressing it, versus filling a balloon with air and try compressing it. With that logic, what goes into the compressor is a gas and what goes out of it is also gas (the coolant gas).

The second thing to remember is that the condenser is a cold cola tin in an eating shop, where water droplets form on the tin. That water is basically gaseous water in the atmosphere turning into liquid which is the definition of condensation.

In the condenser of an air con, the coolant changes state from gas to liquid. Once these two components are remembered, the other two are doing the opposite function. The top of the square is the expansion valve where the coolant in liquid state changes from high pressure liquid to low pressure liquid which is opposite of what a compressor does. On the left of the square is the evaporator which does the opposite of what the condenser does, which is to absorb heat from the atmosphere (the room) and the coolant changes state from liquid to gas.

In the evaporator in a room, cold coolant liquid moves through pipes zig-zag within a small area so it is called a coil. A fan blows atmospheric air from the room onto this coil. So, the reason for the coil is to increase the surface area of the coolant (within the pipe) being blown by the fan. As the fan blows on lots of coolant (within the pipes), it picks up heat from the air of the room and release cool air for humans in the room. This causes the coolant to turn into a gas.

The heated coolant gas is then drawn into a compressor where it is compressed to a high pressure. As a result of being pressurized, the temperature of the coolant increases further.

After coming out of the compressor, the coolant moves to the condenser where it moves in a coil which is blown by a fan. As in the evaporator, the close zig-zag of pipes is made so that the fan blows over a greater surface area of the coolant within the pipes. The coolant at high pressure and high temperature easily releases heat in the condenser and this is emitted to the atmosphere outside the room. As the coolant emits heat, it changes state to a liquid but still under high pressure.

The high-pressure liquid then flows to the expansion valve where it flows through a thin pipe and suddenly moves into a large pipe thereby its pressure is reduced immediately. As pressure decreases the temperature of the liquid lowers even more and it is now ready to pick up more heat.

The cool, low pressure liquid next flows into the coil of the evaporator, most eager to accept heat from the room which is blown onto the coil with a fan.

# Chapter 23

# The Grid

## 23.1 Transmission Lines

In electric power transmission lines, the Ground (Earth) wire is placed on top or higher than the three phase conductors to protect the phase conductors from lightning strikes. This wire is called OHGW (Overhead Ground Wire) or OHEW (Overhead Earth Wire). This OHGW is also joined to all the steel poles which is good since the 'desire' of lightning is only to enter ground, it will find the easiest path to ground and also lightning will always choose the highest point to strike. Therefore, lightning will always strike the OHGW and travel directly to ground. Lightning will not be interested in the phase wires because at both ends of these wires are delta connections of transformers which are not grounded. For single phase LV lines Ground (Earth) is found after the load in the ground rods of homes (or factories) or at the other end in substation transformer whose Star point is grounded. In LV overhead lines, neutral is placed at the very bottom, but in this author's opinion it should be at the very top since in LV overhead lines there is no OHGW. So, if the neutral wire is at the top, it can act like the OHGW wire, draining lightning strike electrons to the ground at star points in transformers or at the other end in ground rods of homes in USA. Only in the U.S. system is neutral and ground wire joined before the home distribution board. But even in the British system, lightning will go to the substation ground which is one of the lowest resistance grounding around, via the neutral wire and not at all be interested in the higher resistance pathway through the loads in homes and then via the green wire to an up to 100 times higher resistance ground rods of homes. In HV, when lightning strikes the OHGW, electrons will continue through the next steel pole and drain to Ground (Earth).

Many are perplexed and are questioning electrical blogs. They ask, in U.S. homes, the ground wire and the neutral wire are joined before the distribution board, why doesn't the high current from neutral (say 10A) go to the ground rod and come up via the green wire to all household equipment's bodies, like the refrigerator's body, the washing machine's body, the oven's body, the rice cooker's body etc. And won't this current in all these equipment's bodies electrocute humans. This can happen because it takes 40V for current to enter a human skin and though neutral is supposed to be 0V, sometimes this author has gotten shocks by touching the N wire. But electrons are just ground (earth) loving, it will **not at all** choose a refrigerator body to go to when it has the ground (earth) to go to. But the problem lies with the connection to the ground rod being broken, then the N which carries as much amps as the L wire is connected to the body of a refrigerator. Still it won't kill a human because N is normally at 0V; an imbalance loading in the three phases can upset this situation and send N to above 40V. But in the U.S. system it is always better to check once in a few years if the connection to ground is alright. This is why electrical laws of all countries calls for an electrical check once in a few years (five years in British law). The advantage of the U.S system over the British system is that if the overhead N wire is somehow broken, normally as a tall truck hits it (since it is currently the lowest wire) the N wire can turn into a L wire on the other side of the load. This will cause L1 and L2 for example to be connected to the loads in the home providing 240 X √3 = 415V going through all household equipment blowing them all up. In the U.S system this won't happen because if connection of N to the substation is broken, there is still the home ground rod acting as the return path for the electrons.

When lightning strikes an OHGW, electrons will just flow to the next steel poles and then to ground and totally not go elsewhere in the grid. This is also similar to a human touching a switchgear that is made live due to a wrong wiring procedure which made the live wire touch the body of the switchgear; this happens especially as wires goes into hole-sawed holes in the body of the switchgear, without proper cable glands. But if a human touch this live switchgear, he will not get electrocuted because the electrons will not be interested in going into his high resistance body and only choose to go to the ground rod via the green wire. For each overhead steel poles, there can be as many as four Ground (Earth) rods through which lighting surges can drain to the ground. But the number of ground rods will depend on the ground resistance at that particular point. The ground resistance of steel poles is specified at $< 10\ \Omega$.

As mentioned in the last paragraph, this author has experienced a shock in touching the neutral wire. This means there is 40V or more in the neutral wire. This could be due to a combination of two factors, the substation is too far away plus the loading in the three phases are not balanced leading to a voltage above 40V in neutral. Note, when the loading (current drawn) in one phase is too high, voltage will drop in that phase, thereby adding all three phase voltages at one moment of time will not be zero volts. Another possibility is that one phase is grounded after the load and is not connected to neutral due to some tampering to steal electricity. If such tampering is done, the combination of voltages in the two remaining phases will not add to zero.

## 23.2 Power Regulator Bank (reactors)

Power regulator banks are located along the line usually in a substation. They are utilized to regulate the voltage on the line to prevent under-voltage or over-voltage conditions. They can be manually or automatically switched on via a thyristor. Reactors are basically coils of varnished wire that introduces inductive reactance ($X_L$=2πfL) into a circuit. As the grid line is switched into a reactor, it acts as a reactance in parallel to the load and therefore reactance reduces (as more resistors are connected in parallel, total resistance goes down) so voltage goes down ($V=IX_L$). This is not immediately logical because all electrical people know that voltage increases with resistance but that is only in series. In parallel as more resistance are connected, voltage goes down so as reactors are switched on, reactance is reduced and therefore voltage reduces. And when a reactor is switched out, or taken out of parallel connection, the total reactance increases and voltage goes up ($V=IX_L$). The equation (64) below and simulation of total resistance on an excel spreadsheet indicates how increasing the number of resistors connected in series reduces total resistance.

$$\frac{1}{R_T}=\frac{1}{R_1}+\frac{1}{R_2}+\frac{1}{R_3}.... \qquad (64)$$

| 1/200,000 | 1/200,000 | 1/200,000 | 1/200,000 | 1/200,000 | 1/200,000 | 1/200,000 |
|---|---|---|---|---|---|---|
| 0.000005 | 0.000005 | 0.000005 | 0.000005 | 0.000005 | 0.000005 | 0.000005 |

.........

This was done till column KQ in excel, meaning connecting 303 resistors of 200,000 Ω each in parallel gives a total resistance of 662Ω.

## 23.3 Shunt Reactors

These coils are connected in parallel with the transformer outgoing windings 275kV or 500kV substation. They are used to compensate for the capacitive var reactance of long underground cables.

## 23.4 Defect of AC lines compared to DC lines

- AC require four conductors L1, L2, L3 (R,Y,B,N) compared to only positive and negative for DC.
- AC lines need to be compensated or else the current wave will not move synchronously with the voltage wave (phase shift occurs). This will lead to more current drawn as shown by the power station output equation below:

$$P = \sqrt{3}VI\cos\theta \quad (65)$$

- If an increasing phase shift occurs, θ increases causing cosθ (or PF) to reduce. P has to remain the same because for example a motor needs 10kW and if the PF is down the I has to increase so it can still get 10kW. But as I increases, the power loss in the overhead lines will increase following equation (66).

$$P_{loss} = I^2R \quad (66)$$

- AC has a skin-effect problem where current only travels on the skin of the conductor due to eddy currents formed in the cable, so the center portion of the cable is not utilized. DC does not have this problem, so the whole surface area of the cable is fully utilized.
- AC also creates eddy current (power wastage) in the cable sheath and armor.

## 23.5 High Voltage (HV) Overhead lines

Common HV used worldwide are: 1000kV, 500KV, 275 KV, 132 KV, 66kV, 33KV, 22kV, 11KV. Stranded conductors are used, because they are stronger. The center of the conductor is steel strands to provide strength. The center of the cable is not used by the AC current anyway because of the skin-effect where current only flows on the skin of the conductor. The central steel of the cable and the skin effects are depicted by Fig. 96 and Fig, 97 respectively.

Fig. 96: Steel reinforcement in the inner radius of overhead HV cables (darker color), the outer radius is aluminum

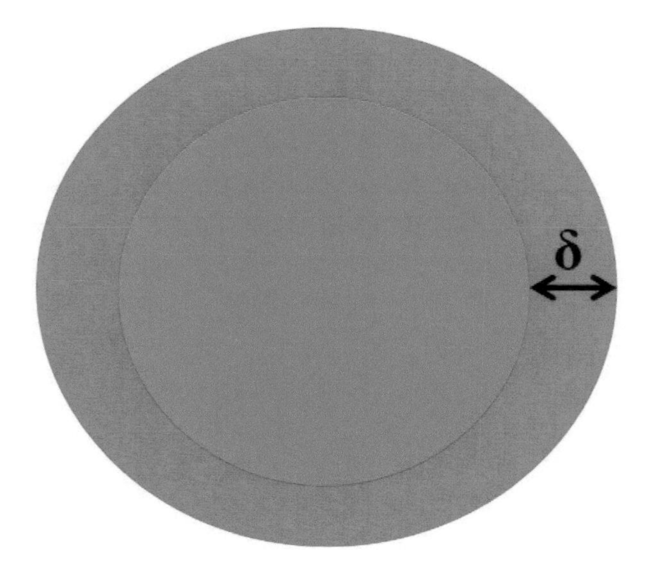


Fig. 97: Skin effect

Fig. 97 indicates the flow of AC in a conductor depicting the skin effect. Current flows mostly in the outer section of the conductor in a region less than a cm thick. The skin depth is termed δ (small letter delta) and increases with frequency of the AC waveform or δ is propotional to the frequency. Cables are run as bundles of two three or four wires together as shown in Fig. 98, this will increase the power carrying capacity of the electric poles. Such cables are officially called bundled cables. The insulators seperating the cables are called spacers and are shown in are shown in Fig. 99. The human hand holding it indicates the actual size.

Electrons in AC is sloshing back and forth. This back and forth sloshing of electrons create a varying magnetic field density. Assume we measure the field density of a piece of magnet moving in a glass tube. The piece of magnet is moved from one end to the other with two strings pulled by a hand or motor. Say the magnet is at the left end, it will have a magnetic Field density measured by a meter, let just say there are 10 field lines above the tube from the magnet, spaced out 1mm from each other. As the magnet is pulled towards the right, another magnetic Field density meter is used to detect the density at the center. But now at time t=1s there will be 10 field lines and at time t=1.01s there will be 10 field lines and t=1.02s there will be 10 field lines. So though there are a fixed 10 field lines above this magnet, at the center there will be 10 spaced out every 0.01s causing a large magnetic Field density. Thereby the magnetic Field density is varying proportionately with the speed of the electron. This is why an electron sloshing back and forth in AC created a varying Field density. Since this electron magnet has the other requirements of FBI; it has F as explained above, it is a magnet so has B with one side of the electron being N and the other S, therefore I will be generated. Using the Fleming's-Right-Hand rule, it can be seen that the current is generated out of the cable. Since each magnetic electron is oriented in random directions, overall a tube-like flow of electron generated is formed where all electrons flow at the outer edge of the wire. This is called eddy current which is the name given to any unplanned or unwanted current generation. This will push out the electrons generated by the power station, so all current in AC cable is flowing only at the outer portion of the cable. Why does this not happen in a DC cable. The DC current can be huge giving a very high magnetic Field density. But the density is not varying so there is no process to generate eddy current in the DC cable or HVDC cable.

For copper skin depth, δ= 8.5 mm (just under 1 cm); meaning electron (current in reverse) only flows from the outer surface of the cable to 8.5 mm inside it, as shown in Fig. 97. At high frequencies, skin depth is smaller. This can be seen in airplanes where the electric generator spins at 400Hz (compared to 50Hz or 60Hz in the grid) resulting in only a very small section near the outer diameter for electrons to flow. To overcome this in airplanes, multiple small cables are run to carry power instead of a single big one. The skin dept surface area of many small cables is higher than an equivalent current carrying capable big cable. With skin effect, the surface area to enable electron flow in many small cables is much higher that a single big cable. Therefore, with AC, it is always better to use many small cables rather than one big diameter cable to carry more current. This author has done this for the incoming of a big hall but no other contractors has seen it (not in regular wiring practice) and it was heatedly debated at the construction site but the owner eventually agreed to let it be. That hall has been running for almost three years now. The reason this method was chosen was that there is a about nine meters of wire needed to be run from the incoming to the main Distribution Board and this route has lots of bends. Four thick cables (16mm$^2$ for L1,L2,L3,N) should not be bent too much because they will be damaged. Four core underground cable also cannot be bent like this. So, the best method is to run four single PVC 2.5mm$^2$ cable per phase. 2.5mm$^2$ cable can carry 21A X 4 = 82A. A single 16mm$^2$ cable can carry 68A. These four 2.5mm$^2$ cables is run in a HDPE (High-density polyethylene) pipe so there are four HDPE pipes. Note regular PVC cables are not designed to handle harsh outside environment but underground cable can handle it. But HDPE pipe which is normally used to carry drinking water is designed to handle harsh outside environment and doesn't leak.

For 33KV, 100mm$^2$ cable is used, the center is made of steel strands as shown in Fig. 96 since current do not flow in the center. This is to provide strength to the cables; the conductors in the circumference is AAAC (All Aluminum Alloy Conductor). For 275KV, 402mm$^2$ is used.

Fig. 98: Bundled cables

Fig. 99: Spaces for bundling cables

When higher amounts of current need to be transmitted, increasing the cross-sectional surface area of the cable is not the answer because of the skin effect. This is why many cables are strung, separated by spaces as shown in Fig. 99. The impedance decreases only slowly with size increase. But size increase has a whole slew of problems; they cost much more, transportation is much more difficult, installation of heavy conductors is much more difficult and the possibility stress, fatigue and eventual breakage due to the weight. A large size cable carrying very high current will also have corona loss which is a radiating purple light due to ionization of air around the conductor. It should be stated that this is the contemporary description of it but Tesla stated that corona is energy emanating from empty space or ether. It is because of all these factors that cables are bundled as two, three, four or six cables with spacers. Spacers insulators must be built strong to overcome the force of heavy wind and magnetic pull during a short circuit.

# Chapter 24

# HVDC Transmission

High voltage direct current (HVDC) systems are used for the following reasons:

1) When the power that need to be transferred is large.
2) When the distance from power stations to consumers is a long.
3) For interconnecting different grids.
4) In big cities DC lines relieve congestion or bypass complicated AC cabling systems.
5) In undersea cables, charging of the capacitor like structure of these cables (three conductors with XLPE in-between making it a longitudinal capacitor) happen six times in one cycle because a high voltage difference in-between phases occurs six times per cycle of one of the phases. This does not happen in HVDC. The undersea cable is also a capacitor between the conductors and the sheath of steel. HVDC need to charge up this capacitor only once upon startup but the HVAC need to charge it up six times per cycle of one of the phases.
6) To enable the use of renewal energy. The output of solar is DC so the output of a large solar farm must use HVDC. For wind turbines, each turbine turns at different speeds so the electric output of one wind turbine cannot be joined to another because there is a phase shift of current from different turbines. But if the output of each wind turbine is converted to DC the output can be joined and transmitted to shore via a HVDC cable.

The conversion of power between AC and DC was enabled with the development of:

1) Mercury-arc valves before the 1970s
2) Thyristors especially Gate Turn-Off (GTO) thyristors after the 1970s.

3) Integrated Gate Commutated Thyristors (IGCTs)
4) MOS Controlled Thyristors (MCTs)
5) Insulated Gate Bipolar Transistors (IGBT)

1 to 4 are CSC (Current Source Converters) or LCC (Line Commutated Converters). (5) is a VSC (Voltage Source Converter). The demand for IGBT (or VSC) has increased dramatically because of the proliferation of off shore wind turbines installations. Imagine a flood control gate controlling the flow of a river. That is how a Gate (G) controls the flow of current from Source (S) to Drain (D) in an IGBT. In bipolar transistors, current must be sent to the Base (Gate) to enable current flow from Collector to Emitter. The names of the many different three-legged transistors are different but they all work with a principle that a small signal at the gate controls bigger current flow between the other two legs. In an IGBT, a voltage at the G induces and electric field which enables current to flow from Collector (C) to the Emitter (E). Thus, the categorization of IGBT as Voltage Source Converter (VSC). If current need to go into the G, it is categorized as Current Source Converters). The demand for various power electronic used in the electric power industry is shown in Fig. 100. Note the IGBT was invented in 1982 and is a combination of the oldest electronic invented, the BJT (Bipolar Junction Transistor) and what is prevalent in all computers and cell phones today, the FET (Field Effect Transistor). As described above the BJT needs current in the controlling Base to allow bigger flows of current between he Collector to the Emitter. But the FET need a field created by a voltage at the Gate to enable much higher current flow between the Source to the Drain. But voltage do not create electric fields, it is current. This is just a categorization of one to be voltage driven and the other current driven. The important fact is that the electrons in the Gate of a FET do not flow to the Drain. Only the electric field of the current flow in the Gate triggers a close circuit between the Collector to the Emitter. BJTs have a higher current carrying capacity while FET's current carrying capacity is small. So, a combination of BJT and FET is made to become the IGBT. So, the names of the legs are also a combination, the computer control signal goes to the Gate (name of leg of FET) and the other two legs are Source (name of leg of BJT) and Drain (name of leg of BJT).

# Power Electronics usage

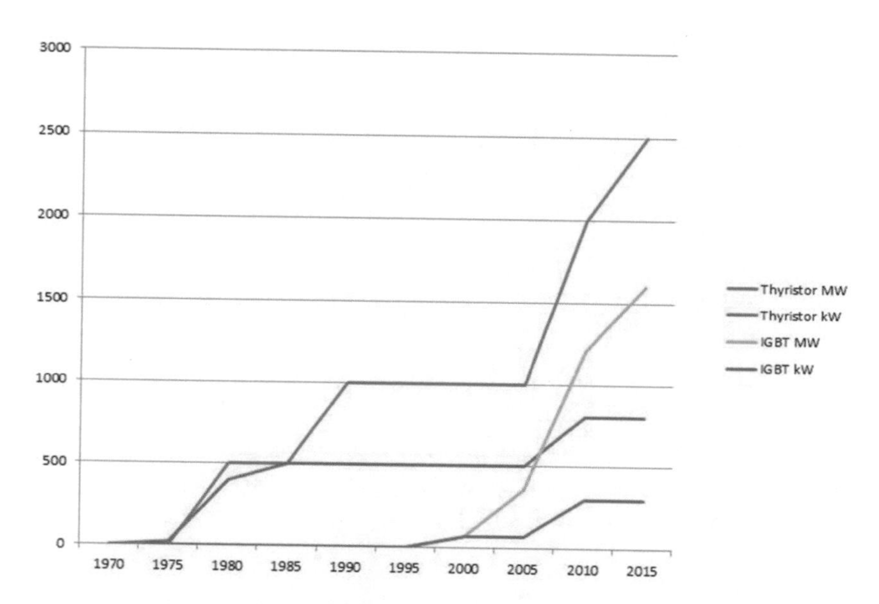


Fig. 100: Demand for various electronics in the electric power industry

Insulation thickness and conductor spacing of an electric cable is determined by the peak voltage. Thus, for an AC cable carrying 275KV, the cable has to be able to withstand voltages of: $\frac{275kV}{0.707} = 389kV$. But a 275kV DC line needs to be able to handle only slightly above 275kV (catering for a slight surge). Which also means a 275kV AC cable can potentially carry 389kV of HVDC.

Due to all these factors, HVDC lines can carry **50% more power** than HVAC lines. The HVDV lines are also **30% cheaper** to construct than equivalent HVAC lines because the towers are simpler to carry fewer cables. The advantages of HVDC over HVAC systems for carrying large power include higher power ratings for a given line and better control of power flows.

The higher carrying capacity of HVDC is important because upgrading current lines or erecting new towers can cost billions of dollars. Many newly installed lines now use HVDC for long distance, high load transmission, especially in developing countries such as China, India and Brazil. In India existing HVAC lines have been used to carry HVDC thereby drastically increasing the current carrying capacity. Each cable can carry 50% more power with HVDC. In a typical HVAC line there is L1, L2, L3 (R, Y, B) bundled cables (two cables per phase) on the right of the tower which leads to 50% X 6 = 300% extra current carrying capacity. This is shown in Fig. 101. If the Single Wire Ground Return (SWGR or Single Wire Earth Return (SWER)) configuration is used there will be another six wires on the left side of the tower which need not be used as neutral, thereby increasing her current carrying capacity to 50% X 12 = 720%.

Fig. 101: HVAC bundled transmission lines on left versus HVDV bundled transmission lines on right. Only positive, negative and Ground for HVDC

## 24.1 Submarine HVDC

As an indication of the current carrying capacity of plastic covered submarine or underground cables versus bare cables, the following statistics are useful:

1) The longest undersea cable (heavily covered by plastic) so far is **580 km, 450kV**, 700MW XLPE HVDC cable between Norway and Netherlands.
2) The longest bare overhead line is the **2,385 km, 600kV**, 7.1GW HVDC Rio Madeira transmission link in Brazil.
3) A close second is the **2,090 km, 800kV**, 7.2GW HVDC Jinping-Sunan transmission link in China.

**Therefore, bare O/H cables can carry 10 X more power 4X further.** Added to the above statistics, underground cables fail much more often than overhead cables and the overall cost of underground cables is up to 400% higher than bare overhead cables, including the electric tower, ceramic insulators and other accessories. Therefore, underground cables are bad compared to overhead lines but HVAC submarine cables are far worse than HVDC submarine cables. Obviously, the main reason overhead bare cables are more capable is the temperature. Putting lots of plastic around cables increases the heat and reduces its current carrying capacity.

Added to this, currently available submarine XLPE cable has lots of capacitance loss. Each conductor is surrounded by a relatively thin insulator (XLPE) and the final layer is a metal sheath. This conductor, the dielectric (XLPE) and the final metal sheath corresponds to a capacitor. Thus, the whole HV submarine cable is a longitudinal capacitor. This capacitance is parallel with the load. When AC is transmitted through the cable, additional current must flow to charge the cable capacitance six times because high voltage difference in-between phases occur six times within the period of one wavelength of one of the phases; this is depicted in Fig. 27. This extra current flow results in energy loss via dissipation of heat; thereby raising the temperature of the submarine cable. This heat eventually weakens the submarine cable and it eventually fails as a current carrying conductor. When DC is used, the cable capacitance is charged up only upon first energizing so the energy loss is less. This is why most undersea cables today are carrying HVDV (High Voltage Direct Current).

Another loss is dielectric loss. Here the varying AC electric field causes small realignment of weakly bonded molecules within the dielectric material causing small vibration of molecules, thus heating up the XLPE. If XLPE is used as the dielectric, this effect becomes significant beyond 63.5 kV in mineral (magnesium oxide) filled cables and 27 kV for non-mineral filled cables. Just to recap, magnesium as school teacher showed burns with a bright white light indicating aggressive reaction when burned. So, it is hard to separate the magnesium with the oxygen. Note metals are conductors because they love to give off electrons so, there are free electrons to move when a voltage is exerted but all metal oxides are insulators to varying degree and magnesium which has a very strong bond with oxygen will tend not to revert to magnesium metal so it is used as one of the best insulators. Another recap is that the only other metal oxide used in electric power is zinc oxide. This is because zinc is one of the least reactive metals which is why it is used to cover up iron as in galvanized steel poles (galvanizing means plating with zinc as a thin outer layer). So, zinc oxide is used to make surge arrestors where a surge will easily dislodge oxygen from the zinc and find a conductive path to earth.

Overall underground power transmission has a significantly higher cost and greater operational limitations but its utilization is expanding rapidly mainly for esthetic reasons. There is also some advantage in right of way. Actually, if Laissez faire business is the only determinant of which designs to choose, all cable will be underground because there is so much more money to be made. Only government can stop wastage of the funds of the people as was done in Thailand where the rich king forbade underground cable and all cables are overhead.

## 24.2 Disadvantage of HVDC

The disadvantages of HVDC are in conversion, switching, control, availability and maintenance. The cost for all these is much higher than for HVAC. With HVAC, the only switching is for switching off or on a line. But for HVDC the switching needs to happen continuously to create the DC. Most of these switching is with thyristors and IGBT. The main circuit breaker for HVDC is much more complicated and was only invented in 2015 by ABB. With this recent invention of HVDC switchgears by ABB, fast switching of HVDC line have been enabled. Switching of HVDC has always been tricky because it does not have the advantage of HVAC where the actual switching can be made to happen when the AC waves cross the zero-volt region. With HVDC, mechanism must be included in the circuit breaker to force current to zero, before the actual switching occurs. Otherwise too much arcing will occur, wearing the silver-plated copper contacts. In HVAC, switching is performed when the AC wave crosses the V=0V so L1, L2, L3 (R, Y, B) cannot be switched OFF or ON at the same time as shown by Fig. 102. The switching of the L1, L2, L3 (R, Y, B) phases must be done about 4ms (5ms British) from each other.

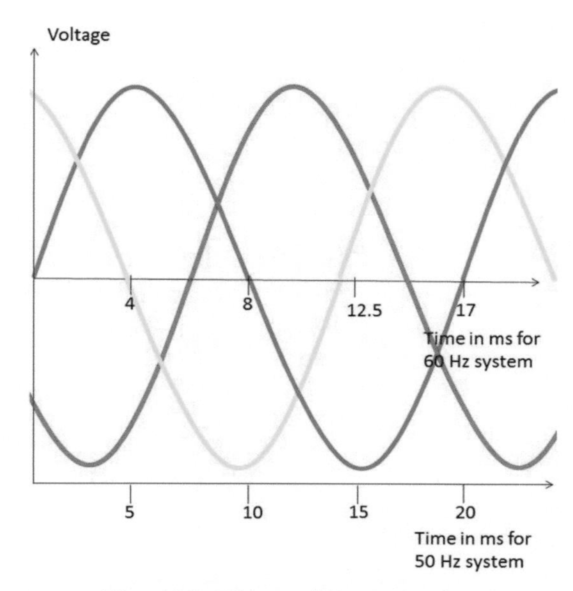


Fig. 102: Three phase waveform

The thyristors and IGBT also induce losses and they have limited overload capacity and are expensive. At smaller transmission distances and loads the losses in these power electronics may be bigger than in an AC transmission line. The cost of the inverters may not be offset by reductions in line construction cost and lower line losses.

# Chapter 25

# Grid Control

## 25.1 Large National or International Grids

The solution for the intermittant nature of renewal power is to join up grids as much as possible. Thus a low wind flow over wind turbines in South Dakota can be backed up by gas turbines in Texas or solar farms in California.

So, North America has decided to join up it's nine major grids with Back-To-Back DC lines (B2B). These B2B can be the length of two classrooms to convert AC to DC at one end and DC back to AC at the other end. A large substation is being planned in the U.S. called Tres Amigas which will interconnect the three large grids in the U.S, the Western Interconnection, the Eastern Interconnection and the Electric Reliability Council of Texas (ERCOT) grid. Electricity from each grid is converted to DC and flows in a smaller DC high-temperature superconductor (HTS) wires in a 58 $km^2$ area where any of the three grids can pull out power or supply intermittent renewal energy from wind or solar farms. Superconductors have the ability to keep electrons flowing within them for long periods (up to 100,000 years) of time making them energy storage devices. The Tres Amigas substation can initially handle 5 GW but will be expanded to 30 GW. HTS wires conduct electricity about 200 times better than copper wires of the same dimensions.

Europe is taking another route, where the International Electrotechnical Commission (IEC) did research to determine how extensive usage of renewal energy can be achieved to prevent global warming and reduce dependence on fossil fuels. Their proposed design is a Europe wide grid to cater to the intermittent sources of electric power like wind or solar. This way, extra power from fossil fuel, nuclear power stations or other renewal energy sources can back-up a sudden low wind situation around wind farms or sudden cloud-cover over solar farms. Therefore, the Synchronous Grid of Continental Europe was built and it is currently the largest electrical grid in the world. The frequency of the grid is 50Hz and it serves over 400 million customers in 24 countries. This way when there is low wind over off-shore wind farms of U.K., power can be drawn from large solar farms in Spain. This author is located in South East Asia which consist of islands and peninsulas, so he is working on a research to suspend electric cables within submarine pipes to enable a South East Asian grid. This way, when cloud hovers over a large solar farm in Malaysia, geothermal energy from Indonesia can back it up.

Large energy storage has also been determined to be one of the solutions; batteries with liquid electrodes have been suggested as a form of grid energy storage. In is obvious from our cell phones and laptops that battery storage has improved remarkably over recent times. But a sure method which do not need waiting too long for better batteries is already implemented in Germany where solar panels powered pumps, will pump water up a high tank during the day and at night water is released from these tanks to produce hydroelectricity.

It should also be noted that computer simulation have shown that giant faults cannot pass a B2B system, a fault cannot bring down the whole North American power system but it may be possible that a major fault can bring down the whole European grid.

## 25.2 Control of Grids

In the vast grids, the V and I tend to not flow synchronously. This will cause the PF to decrease, causing I to increase and thereby power loss to increase due to the formula $P_{loss} = I^2R$. To control the drifting of the I wave from the V wave, the following are used to produce vars (cause leading PF) which means pushing the current wave forward so that it goes back to be in phase with the voltage wave. This is the greatest requirement in grids because loads all over the country are mostly coils (motors) which causes lagging PF or current wave to be slower than the voltage wave. Other than lagging loads there are resistive loads which do not affect the PF. At least this author has never seen a load that causes leading PF. Equipment installed by power companies in the grid that causes leading PF are called compensators:

1) Generators run as synchronous motors (synchro mode) and over-exited with extra DC current in the stator solenoid – **remember as more DC => Leading.**
2) Capacitor banks (Fig. 105).

3) The capacitance of underground cables.
4) The little capacitance of overhead lines.
5) SVR (Static var Regulator).
   Reactors (Fig. 104) opened or disconnected from the overhead lines because reactors are basically, coils and reduction of coils will cause leading PF
6) Production of vars (increase capacitance) is by decreasing loads on lines (load shedding).

To consume vars (or cause lagging PF or to push the current wave back so that it will be in phase with the voltage) the following equipment are used:

1) Induction motors.
2) All other coil-based loads.
3) Generators run as synchronous motors under-excited – remembered as **less DC => Lagging.**
4) Reactors in substations being switched on to the grid line, these are shown in Fig. 103.
5) Inductance of overhead lines.
6) The little inductance of underground cables.
7) Transformer inductances.
8) Line commutated static converters.

Number (1) in the first list above is done by playing around with the DC going into the rotor of the generator. Slightly higher than optimal DC (optimal DC is when V and I are in phase) will cause the voltage wave to move slower because it will be harder to move the rotor because it has become a stronger electromagnet. Note that the rotor movement is exactly proportional to the voltage wave in the grid in steam turbine power stations and multiples of this in other power stations following equation (22). This will cause the voltage wave to move slower and eventually be in phase with the originally slower current wave; this is called var absorbing.

## 25.3 Inductor, Capacitor, Resistor model of HV overhead lines

Overhead lines are modeled as a resistors and inductors (due to magnetic field around the cables) in series with capacitors shunting the overhead lines to ground. Capacitance is negligible for overhead lines spans shorter than 50 miles (80 km). Transmission line are both inductive (absorb vars) and capacitive (generate vars); meaning they absorbs and generates reactive power respectively. Under light load, power lines are more capacitive (generates vars) than it consumes (inductive). When there is a large loading in the power lines it consumes and generates the same amount of reactive power. But under heavy loads, the power lines absorb more reactive power than it generates.

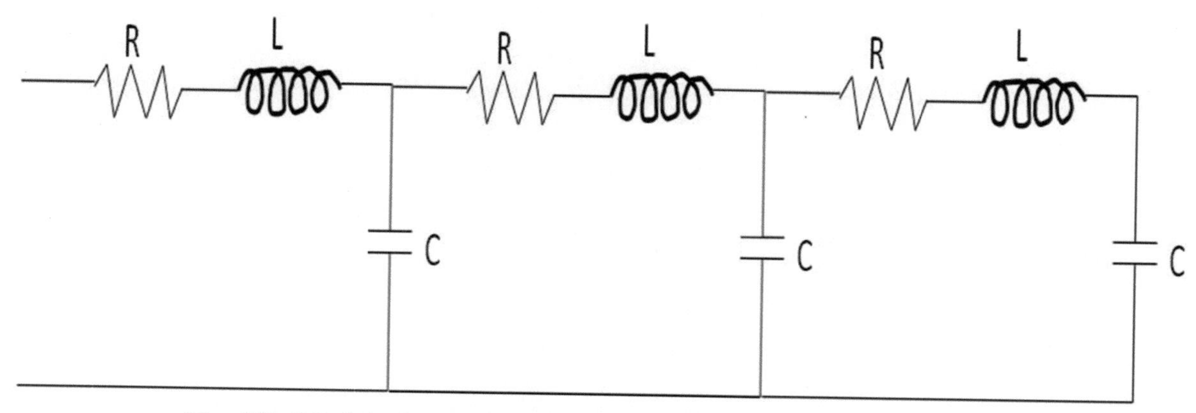


Fig. 103: Model of transmission line which depicts it as resistors in series with inductors with capacitors shunting to ground

HV lines are spaced further apart to achieve sufficient insulation in-between them so they have higher inductance. What happens is when lines are further apart; the field lines do not cancel out each other (superimpose with each other) as when they are closer together as in a three core, three phase underground cable. The field lines do not cancel and there is no insulator to reduce the flow of field lines which are not superimposed. Upon observing from one point of L1 conductor, the field lines will cut a large length of L2 conductor with increasing concentric radius of field lines. So effectively L2 conductor will experience more field lines compared to if the distance between L1 and L2 is smaller. Since the number of field lines is proportional to the inductance, the inductance is increased with increasing distance between L1 and L2.

For underground cables the field lines cancel out each other so the inductance in series is much smaller. Also, the space in-between the phase conductors are very small resulting in much smaller number of field lines crossing from one point of L1 to any length of L2. In addition, only a few field lines that can pass through the XLPE in-between the power lines. But the capacitance in parallel is relatively very high because the three phase conductors with XLPE in-between is equal to a longitudinal capacitor.

For overhead lines, the greater the radius of the power lines, the lower it's inductance. But large radius power lines are hard to work with due to their inflexibility and weight. Thus, two or more conductors are bundled together to approximate a large diameter conductor. This will also reduce radiative corona loss.

Since a voltage V is applied to a pair of conductors separated by a dielectric (air), charges q of equal magnitude but opposite sign will accumulate on the conductors. Capacitance C between the two conductors is defined by:

$$C = \frac{a}{v} \quad (67)$$

The capacitance of a single-phase transmission line is given by:

$$C = \frac{2\pi\varepsilon}{\ln(\frac{D}{r})} F/m \quad (68)$$

Where $\varepsilon = 8.85 \times 10^{-12}$ F/m

This capacitance is only significant beyond 80 KM.

**Without balancing watt (active power) between source and load, the frequency in the system will be affected. Without balancing vars (ensuring PF ≈ 1), voltage in the system will be affected.**

The production of reactive power should be as close as possible to the inductive loads because vars cannot be transmitted far over the line since $X \gg R$ in a grid. Factories and big buildings are required to have capacitor banks to produce vars to balance the vars consumed by the coils in their induction motors and other loads. Note humans only use inductive machines (motors etc.). Capacitors are used to balance the inductive load.

The further a city is from the power station the more V and I wave move apart, meaning the grid at that point is consuming more vars. Thus compensators (1-7 at the beginning of this chapter) need to be placed near the load or close to the city or industrial estate to ensure PF≈1. If microprocessor-controlled thyristors are used to control the current flowing through a reactor-capacitor combination, a continuous control of the reactive power can be achieved; such a device is called SVC (Static Var Compensator).

For example, for the city this author lives in, previously there was enough power generation in the city but currently the power generation in the city is only about a third of the demand. Most of the power is generated about 1000 miles away in a hydroelectric power station. Thus, the phase shifting of V and I is quite high in the city. So, two 26MW hydroelectric generators (out of four) about 100 miles away is run in synchro mode to compensate, but recently even this is insufficient so a 275kV capacitor bank is installed just outside the city (Fig. 105).

Fig. 104: Reactor Banks at a Substation, 33kV on left and 275kV on right

Fig. 105: Capacitor banks in Substation, 275 kV

## 25.4 Power Quality

Though the word power quality seems to indicate electrical power but it is the voltage variation which need to be solved because the current demand by appliances cannot be controlled. Various categories of power quality defects are termed as swells, dips or sag, random, spikes or surges, under voltage, over voltage, brownouts, and harmonic.

Power quality has traditionally been affected by switching on or off large loads on the grid. Note that switching off or breaking HV contacts is two times more dangerous than switching on or closing HV contacts. of loads in the grid. A large factory switching on or off a big equipment will affect the whole grid. Starting large generators also affect the power quality, which are generally solved by AVR (Automatic Voltage Regulators) and other equipment at the power station.

Power quality has become important recently with the increase in sensitive electrical devices being built. Switch Mode Power Supplies (SMPS) has become the standard intake for all electrical devices from hand phones, televisions all the way to advanced equipment in factories. SMPS works with the principle of initial rectification of the power company's 60 Hz (50 Hz British) electrical waves with a bridge diode. The DC produced is then inverted into an ever-increasing (over the years) frequency AC wave (up to 1,000,000 Hz). This is then stepped down with a transformer and rectified again to be used by devices. This seems to be an optimum solution. High frequency AC rectifies to a very flat DC and the transformer decreases in size with frequency, thereby saving copper. But there is a defect, if the system above is fed with an initial uptake of not perfect power company's 60 Hz (50 Hz British) but spiky waves with lots of noise or harmonics. The initial rectification system which is supposed to get rid of all variations does not clean the 'dirty' waves which do get through the initial rectification thereby destroying expensive equipment. For example, a high spike on the positive side will enter the bridge diode capacitor system of the rectifier and remain a high positive all through the SMPS as a high frequency spike, thereby going into the load. Doctors in developing countries have long complained that expensive medical equipment imported from overseas tends to get damaged very fast. The same must be the case for all sensitive equipment users in industry of developing countries. This is because in more advanced countries many systems have been implemented to ensure the AC wave is as noise free as possible. In the power company where this author worked, engineers know about the latest devices developed to enhance power quality but they cannot install it because there is only one grid line running across the whole state and to install those devices require downing the line which the politicians will not agree to.

The perfect solution developed today is to use a UPS (uninterruptible power supply). In this case the power company's 60 Hz (50 Hz British) is stored in a battery and output from this battery is inverted into high quality 60 Hz (50 Hz British) supply. This is because only battery can supply a perfect DC.

Another solution to power quality problems is to use a phase shifting transformers which have a non-integer or complex turn ratio such as 1: $ej^{30}$. This can cause a phase shift of voltage and thereby absorb or produce reactance. Below are three more problems power quality problem causes and solutions to them.

Resonance between the capacitors and inductors in the grid. This can cause a high voltage surge which can breakdown insulation. Inductive-capacitive-inductive (LCL)-type line filters are the normal solution for this problem in the grid.

Travelling waves are a problem especially when high voltage waves reflect on reaching an end or junction of the grid. The affect is that these reflected waved get superimposed with the original AC wave creating a wave that has an amplitude that is several times the original AC wave. The solution to this problem is to have a proper switching sequence. A smarter grid that can detect any surge in amplitude is a fool proof solution.

Reduction in sending HV over underground cable will help because the likelihood of air insulation breaking down for overhead cable is much less.

Previously it was written that **without balancing watt (active power) between source and load, the frequency in the system will be affected. Without balancing vars (ensuring PF ≈ 1), voltage in the system will be affected.** How this happens is that as the demand is higher than the supply, the immediate affect is that more current is drawn from the stator coils of the generators. But as this happens, the stator coils become more magnetic which cause the rotor to turn slower. This will cause frequency to reduce. But as this demand is sustained to be higher than supply, voltage will drop also. This is s general rule of electrical as demand goes too high voltage will drop. One story to tell is that this author was doing some wiring in business area on Christmas day. All the shops were closed and the voltage was 260V while the normal voltage is 240V. The low demand shoots up the voltage. Conversely a high demand will cause voltage to drop. Similarly, the immediate effect of not balancing inductive and capacitive vars is to affect voltage but a sustained one will cause current to go up in the stator coils of the generators and slow it down affecting frequency as well.

Such a scenario happens due to poor voltage control at the control room. Smarter grid to detect and eventually automatically control in the control room is a solution to this problem. The control room staff cannot increase the load but they can reduce the supply by instructing power stations to shut down generators. A gas turbine power station can have about eight generators; instruction can be given to shut down one or two of these generators. Such methods of course take time for example starting up a GT takes 45 minutes. Other methods of control in the control room are the personnel's hands are switching on or off reactors, large capacitor banks or instructing hydro dams to shut down one or more of the penstock pipes and running the generators below them as synchronous motors (synchro mode). The worst-case scenario would be to do load shedding to reduce demand so as to balance supply and demand of power. This have to be done sometimes to save the rest of the grid. In the latter case there is simply no option to start new generator because all has already been started up or no new generator can start at that moment of time. So, in such a case the load shedding is done and it is usually the homes that are cut off from power supply because they are the least critical. Usually the highest priority is to have electric power is given to hospitals, number two are the political and military institutions and then factories and lastly homes. Shutting down factories usually involve payment for their loss of power in terms of their loss of production time plus all the chemicals damaged as in high-tech chip manufacturing. Therefore it may be expensive for an electric utility to shut down some factories. Why is there a case of a generator not being able to be started at exactly a particular moment of time is because duration between the time instruction is given by a control room personnel to the time the generator sends energy to the grid varies for different types of generator. For coal plants it can be as long as eight hours, for ICE it can be over an hour, for a GT it can be about 45 minutes and for a hydro, it can be less than 30 seconds. Basically, a control room personnel is like a bus driver who has to continuously monitor conditions and take actions. Recently software determines generator starting instructions but he control room personnel still has override ability. This author was a control room personnel and the room was called the State Dispatch Center (SDC) but the team working in the room called it the Stress Development Center.

## 25.5 Smart Grids

Smart grids are grids with intelligence built in. They are optimally used to manage multiple sources of renewal energy in a small region. But increasingly they will be fully employed in the whole grid. Smart meters have also been deployed to a high degree in USA. These meters can send and receive information from utilities and communicate wirelessly with appliances especially the cloth washing machine which is one of the biggest consumers of electric power in U.S. homes especially because of their dryers. A customer can switch on the washing machine but it will not switch on till the load demand in the region is low; normally around 2-3am. Eventually more and more of the appliances in the homes can be monitored which gives the power company an idea of how much inductive or capacitive load are out there. Electric meter readings will be sent directly to the power stations. There are also developments to use the internet protocol (TCP/IP) which is very much faster than the currently used Ethernet (mostly with SCADA). Currently most power companies are worried about putting all their equipment on the internet because of the possible hacker attack but the Ethernet can just as easily be hacked. As virus protection develops software is developed, the firewall can be so high that it will be a Himalayan task to launch an attack. Gone are the days when an individual can launch an attack, today it takes a large country to launch a virus attack. With the TCP/IP carrying all the faults, response time can be faster for sudden faults. Sudden surge in load or insufficient generation can be rectified faster. Generators can be started automatically instead of the manual system currently practiced. Even load shedding can be done automatically saving the decision-making time of the control room personnel. Basically, the Smart Grid utilizes the advances made in computer and communications technology into the grid. Among the various devices developed are sensors with built-in microprocessors which can take actions instead of just detecting faults. These intelligent sensors are the eventual goal. Say, on a rural line, a fault is detected, if the normal route is taken, this fault signal is sent via fiber optic cables to the central control room and an automatic or manual reaction is taken. But this reaction delay time may be too long. It would be optimum if the sensor can take actions by itself thus preventing a small fault from escalating into a large fault.

GE is working on a digital twin for all future gas turbines it produces. With sensors installed in all future GE GTs, when a bearing in a GE GT in Malaysia fails, it will show up in a computer running in a cloud in the USA. This way only that particular bearing needs to be changed while the current method is to change all bearings on a fixed Preventive Maintenance (PM) schedule. Having worked in the Western Digital (WD) factory (after working in the electric utility), this author was slightly heart broken by the good bearings which are changed out every PM, though most of them are perfectly good. In the WD factory the PM schedule is arbitrarily made because a study to determine when the bearings will fail will take too long. In WD, machines and processes change every day (quite different from an electric utility) so there is little time to make studies on mechanical parts. The focus is only on the ever-increasing quality and data density of the hard disk produced. But with the new GE system bearing could be changed out only after two years on or more on a GT. This will save expenses in spare parts and in down time of the generator. Note, every part of a GT is of exceptional quality so they are really expensive. Parts like the fins in the GT are actually grown as in a crystallization process instead of molded or milled which explains its high price. Eventually these sensors will be installed in all machines in all industries as a standard, each with an IP address so they can be accessed or even take some control functions. This is the Internet of Things (IoT) which most think tanks agree will be the wave of the future. This will enable efficiency in resource usage but some are saying it can also be used in the, "Big Brother Scenario". Basically, it all boils down to human decency and that must be taught to children from a young age. A knife can be used to cut vegetables but it can also be used to kill another human. If there is morality, any device of low or high tech can be utilized for the good of humanity if humans are on the whole moral beings.

# ABOUT THE AUTHOR

Dr. Prashobh Karunakaran is a Senior Lecturer in Electrical Engineering at University College of Technology Sarawak (UCTS), Malaysia. He is a professional engineer and also runs an electrical consulting and contracting business together with his electrical technical training school. He did his Bachelors and Masters at South Dakota State University, SD, USA and his PhD at Universiti Malaysia Sarawak (UNIMAS). His wife, Sreeja is also an electrical engineer and they have three children, Prashanth, Shanthi and Arjun of whom Prashanth and Shanthi are currently pursuing their electrical engineering degrees.

Printed in the United States
By Bookmasters